Everyday Physics

Colors, Light and Optical Illusions

Everyday Physics

Colors, Light and Optical Illusions

Michel A. Van Hove

W⦵ World Scientific

NEW JERSEY · LONDON · SINGAPORE · BEIJING · SHANGHAI · HONG KONG · TAIPEI · CHENNAI · TOKYO

Published by

World Scientific Publishing Co. Pte. Ltd.

5 Toh Tuck Link, Singapore 596224

USA office: 27 Warren Street, Suite 401-402, Hackensack, NJ 07601

UK office: 57 Shelton Street, Covent Garden, London WC2H 9HE

Library of Congress Cataloging-in-Publication Data
Names: Hove, Michel A. Van, author.
Title: Everyday physics : colors, light and optical illusions / Michel A. Van Hove.
Description: New Jersey : World Scientific, [2022] | Includes bibliographical references and index.
Identifiers: LCCN 2021021109 | ISBN 9789811238338 (hardcover) |
   ISBN 9789811239311 (paperback) | ISBN 9789811238345 (ebook) |
   ISBN 9789811238352 (ebook other)
Subjects: LCSH: Physics.
Classification: LCC QC23 .H8818 2022 | DDC 535--dc23
LC record available at https://lccn.loc.gov/2021021109

British Library Cataloguing-in-Publication Data
A catalogue record for this book is available from the British Library.

The source and copyright holder of all Figures is The Author, unless otherwise stated.

Copyright © 2022 by World Scientific Publishing Co. Pte. Ltd.

For any available supplementary material, please visit
https://www.worldscientific.com/worldscibooks/10.1142/12316#t=suppl

Dedicated
to the late
Léon Van Hove
promotor of
scientific
popularization

Preface

This book welcomes all those who are curious about nature and have a desire to understand how nature works. The emphasis is on conceptual ideas of physics and their interconnections, while avoiding mathematics entirely. The approach is to explore intriguing topics by asking and discussing questions: thereby the reader can participate in developing answers, which enables a more satisfactory and long-term understanding than is achievable with memorization.

This book grew out of a popular General Education course that I gave for six years at two universities in Hong Kong, entitled *"Everyday Physics for Future Executives"*. It was inspired by a very successful course given for many years at the University of California at Berkeley by Professor Richard A. Muller, initially entitled *"Physics for Future Presidents"*, and available in book form.[1] As Professor Muller wrote: *"Can physics be taught without math? Of course! Math is a tool for computation, but it is not the essence of physics."*

---

[1] Richard A. Muller, *"Physics and Technology for Future Presidents: An Introduction to the Essential Physics Every World Leader Needs to Know"*, Princeton University Press, 2010, ISBN-13: 9780691135045.

The approach of this book recognizes that math is a hurdle which frightens many people away from physics. Actually, by avoiding math altogether, everyone, regardless of prior knowledge, can <u>feel</u> and enjoy physics (and other sciences), thereby understanding a multitude of daily experiences, from colors to nuclear energy. That is the principle of this book. While it shows a few sums, subtractions and multiplications to illustrate what math can contribute, these simple calculations are <u>not</u> at all needed to understand the physics. Math is needed for rigor, precision and prediction, but not for understanding the essence of physics.

I chose for this volume the topic of "Colors, light and optical illusions", because we face colors and light every waking minute of our lives, and we experience optical illusions more often than we realize.

First and foremost, I acknowledge both the City University of Hong Kong and the Hong Kong Baptist University for the opportunity to develop and fine-tune the abovementioned General Education course, open to all students at those universities. I was most fortunate to benefit from incisive discussions and substantive assistance from Klaus E. Hermann (Berlin), including ideas, graphics and detailed scrutiny. I also benefited from very valuable suggestions from Nai-Ho Cheung and Sheng-Wei Wang (Hong Kong), as well as from Han Boxma and Willem Maat (the Netherlands). Of particular value was advice from Robert G. Greenler (Wisconsin). All remaining errors are mine and mine alone.

Michel A. Van Hove

Emeritus Chair Professor, Hong Kong Baptist University

Retired Chair Professor, City University of Hong Kong

Retired Senior Scientist, Lawrence Berkeley National Laboratory, University of California

Hong Kong, June 2021

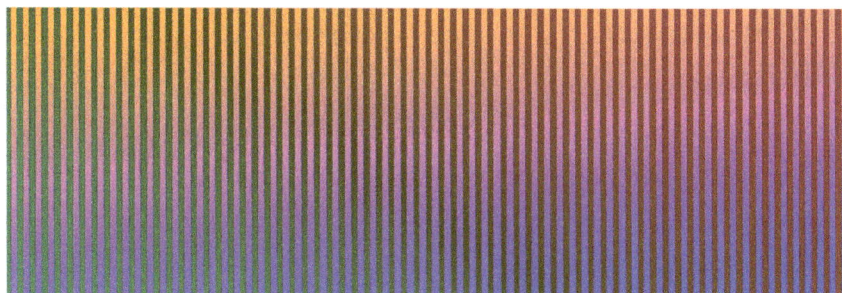

Contents

Animations and Cutout Models are Available to Buyers of this Book at https://worldscientific.com/worldscibooks/10.1142/12316#t=suppl, also reachable via the following QR code (see more details on page 255).

List of Concepts, Connections and Terminology

Numbers refer to Chapters and Sections

additive process (RGB) 4.1: The RGB color system is based on red, green and blue primary colors and adds these three colors together, starting from no light (black). This contrasts with the subtractive process used in the CMY color system, which subtracts cyan, magenta and blue from white. (See also **subtractive process (CMY)**.)

black 2.5, 4.1: Black is the absence of light (or at least the near-total absence of light, in which case it can be called dark gray). A computer or television screen that is turned off is black if no other light shines on it. A painted or printed page is black if there is enough black paint or ink to absorb all light that may shine on it. (See also **colors**.)

brain 2, 3, 5.1, 8, 10: In this book, the brain includes the nervous system (neurons) leading from the eyes to the brain. Together they process the electrical signals sent by the cones and rods in the eyes' retina, to produce a mental image that we "see" in our mind (the mental image may be substantially different from the physical image on the retina: for example, the brain can compensate our color perception for lighting with a non-white color, but it can also create optical illusions). (See also **vision** and **optical illusions**.)

brightness 2.2: Brightness (also called intensity, luminance or radiance) is the strength of light: for example, sunlight is much brighter than a weak lamp producing the same white color. (See also **color**, **hue**, and **saturation**.)

CMY color system 4.1, 6.1: The CMY color system combines the three colors cyan (C), magenta (M) and yellow (Y) to compose many visible colors. It is used primarily for printing. When black is added as a fourth ink color, this system is called CMYK. (See also **colors**, **HSL color system**, and **RGB color system**.)

color 2, 4: Color is a property of light that our eyes normally can detect. Color can be categorized by hue (such as red, yellow, green, cyan, blue, magenta, white or black) and by brightness (also called intensity, luminance or radiance). The word "color" can be taken strictly as "hue only" (ignoring the light's brightness) or "hue and brightness" (for example, yellow and brown have the same hue but different intensities but can be called different colors). Any given color is a particular combination of hues, for example magenta and purple are different combinations of red and blue, while gray and white are equal combinations of red, green and blue. Color is detected by cones (in the retina of eyes), by light-sensitive chemicals (in photographic film) or by electronic detectors (in digital cameras). Color blindness and color vision deficiency in eyes can reduce sensitivity to certain colors. Light is created by various physical or chemical processes, such as nuclear reactions in the Sun, chemical explosions, fire, lasers, lightning, electroluminescence (in electronic displays), *etc.* Most objects reflect light and change the color of the reflected light by modifying the relative amount of each hue (such as removing blue light from white light, leaving yellow light).

color blindness 8: Color blindness is an extreme form of color vision deficiency, which is a weakness of sensitivity of the eyes' cones to certain colors. Strictly, color blindness refers to total insensitivity to light by "red", "green" or "blue" cones. For example, the "green" cones could be insensitive to all light; however, the "red" or "blue" cones may then still be somewhat sensitive to green light.

color vision deficiency (also called color deficiency) 8: See **color blind-ness**.

combination of pure colors 2.5: Most light is composed of multiple pure colors (that is, different hues). For example, magenta is composed of pure red and pure blue. (See also **pure colors**.)

complementary colors 2.7, 4.2, 10.1.1: Pairs of complementary colors have a strong contrast in hue (for equal brightness). For example, red *versus* cyan, or blue *versus* yellow. Complementary colors add up pairwise to white when mixed: thus, red mixed with cyan (in the correct ratio) produces white. (See also **colors**.)

cones 2.4, 3, 5.1.1, 5.1.2: The cones are molecular detectors of light in the eye's retina (their name comes from their conical shape). They normally are predominantly sensitive to either red light ("red" or "L" cones), or green light ("green" or "M" cones), or blue light ("blue" or "S" cones). The cones are also somewhat sensitive to the two remaining colors (for example, the "red" cones also sense green and blue light). The retina also has other detectors, rods, that are more sensitive to weak light than to color. (See also **rods**.)

creating color 6: Color can be composed by mixing pure colors. For example, in electronic displays, the colors red (R), green (G) and blue (B) are mixed; in printing, the colors cyan (C), magenta (M) and yellow (Y) are mixed; in painting, the artist also mixes colors on a palette. (See also **displaying colors.**)

cycle of imaging 3: In photography and moviemaking, many transfor-mations occur between the original scene being photographed and the final scene perceived in our minds. These transformations include collecting light, recording and storing images, editing and displaying them, and interpreting them in the brain. The resulting mental images often look rather different from the original scene in various ways (color, sharpness, 3D structure, *etc.*).

depth information 10.3: In normal vision, our brain must reconstruct a 3D mental image out of a 2D physical image on our eyes' retina. Going from 2D to 3D needs depth information, namely distances to objects in the scene, which are not available in a 2D image. The brain guesses depth information from many clues (such as stereoscopic differences seen by our two eyes) and experience (such as knowledge of the sizes of familiar objects). A wrong guess can lead to optical illusions, for example. (See also **vision** and **optical illusions**.)

detecting light and color 3, 5: Eyes and cameras use lenses to focus light on the retina or detectors for sharp vision. Eyes detect colors using its "red", "green" and "blue" cones; weak light is detected with our eyes' rods; the resulting electrical signals are sent to our brain for processing. Film photography uses chemical processes with light- and color-sensitive materials to produce a physical picture on paper or film. Digital photography also uses light- and color-sensitive materials: these produce electrical signals that are collected and stored digitally in a computer. (See also **recording images**.)

diffuse reflection 4.2: Reflection of light is diffuse when the light is reflected in many directions. Diffuse reflection contrasts with specular reflection, in which light is reflected in a single direction as by a mirror. (See also **specular reflection**.)

digital photography 5.3: Digital photography uses light- and color-sensitive materials which produce electrical signals that are collected and stored digitally in a computer. Advantages over chemical film photography include: fast recording, loss-free long-term compact storage, shipment over the internet, easy multiple copying, and limitless editing in graphical software. (See also **film photography**.)

displaying color 3, 6: Color is displayed as paint or ink on paper and film by mixing paints or tiny ink dots. On electronic screens in television and computing, color is displayed with tiny electronic emitters producing three different colors. (See also **creating color**.)

electromagnetic radiation 2.1: Electromagnetic radiation is a kind of wave that mixes electric and magnetic components. It includes all light, as well as, for example, radio waves and x-rays. (See also **electromagnetic spectrum**.)

electromagnetic spectrum 2.5, 7.1: Sunlight and any other light can be split up into the solar spectrum (for example with a glass prism). The solar spectrum is the visible part of the electromagnetic spectrum, which extends beyond the solar spectrum into infrared light, radio waves, ultraviolet light, x-rays, *etc*. Light in the electromagnetic spectrum consists of waves with a range of color-dependent wavelengths and frequencies. (See also **solar spectrum**, **electromagnetic radiation** and **waves**.)

electronic displays 6.2, 6.3: Electronic displays include television screens, computer monitors, and smartphone displays. They consist of a grid of tiny light emitters (pixels) that can produce red, green or blue light of different intensities (brightnesses): the mix of red, green and blue determines the particular color emitted by an emitter, following the RGB color system.

electronic recording 3, 5.3: An image is focused by lenses onto a grid of tiny electronic detectors that are sensitive to red, green or blue. The electric signals are converted to RGB intensities for each pixel (picture element) of the image and stored for later viewing.

eye 2, 5.1: Human (and many animal) eyes contain, among other components: the cornea (an external transparent layer that focuses light) covering the iris and pupil (an adjustable circular "curtain" with hole); the lens (that adjusts the focusing of light for objects at different distances); and the retina (with light-sensitive cones and rods) with a fovea (a dense collection of detectors for sharp vision).

film photography 3, 5.2: Film photography uses chemical processes with light- and color-sensitive materials to produce a physical picture on paper or film. Disadvantages over digital photography include: slow chemical processing (less so in instant photography), degradation of image over time, difficult copying and editing, no shipment over the internet. (See also **digital photography**.)

focusing 5.1.3: Light entering eyes and camera for detection must be focused to produce sharp images. The cornea and the lens do the focusing.

fovea 5.1.2: See **eye**.

frequency (of light or wave) 2.5: See **waves**.

gamut 6.2.2: A range of colors, for example the RGB triangle or the CIE 1931 color space chromaticity diagram.

HSL color system 4.1: The HSL color system organizes colors by hue (H), saturation (S) and luminance (L). Luminance is similar to brightness. (See also **colors, hue, saturation, brightness, CMY color system**, and **RGB color system**.)

hue 2.2, 2.3, 4.1, 6.2.3, 7.6: Hue specifies color without regard to brightness (intensity). Thus, bright and dark red have the same hue, despite different intensities. The main hues include red, yellow, green, cyan, blue and magenta. The hues correspond to the "pure" colors of the solar spectrum, ignoring their intensities. (See also **colors, saturation**, and **brightness**.)

human vision 5.1: Human vision consists of the eyes, the nerves (neurons) connecting the eyes to the brain, and parts of the brain itself. Each of these three components processes the observed images, resulting in a mental image that may be rather different from the observed image (for example, in terms of adjustments in color and contrast, and leading sometimes to optical illusions). The mental image is largely and immediately forgotten. (See also **vision** and **optical illusions**.)

imaging vs. imagining 10.3.1: Vision includes two major steps: 2D physical imaging, and 3D mental imagining. Imaging is a physical process that converts a scene to a 2D image. This 2D image is then processed by the brain, for example to adjust colors and contrasts, and to add depth information to construct an imagined 3D mental image (which may lead to optical illusions). (See also **vision** and **optical illusions**.)

infrared light (IR) 2.5, 5.1.4, 7.1, 7.4: The solar spectrum extends beyond the range of visible light, which spans from red (at lower frequencies) to blue (at higher frequencies). Frequencies below red include invisible infrared light, radio waves, *etc.* (Frequencies above blue include invisible ultraviolet light, x-rays, *etc.*) (See also **electromagnetic spectrum, waves** and **ultraviolet light**.)

ink color 6.1: Printing in color on paper mainly uses the CMY (or CMYK) color system, for example in inkjet and laser printers. Inkjet printers deposit tiny droplets of cyan, magenta and yellow ink, as well as black droplets. The frequent inclusion of black ink leads to the name CMYK, where K stands for black. (See also **colors** and **paint color**.)

intensity 2.2, 4.3: See **brightness**.

iris of eye 5.1.3: See **eye**.

L cones 8.2: See **cones**.

laser 2.5, 6.3: A laser (which name is abbreviated from "light amplification by stimulated emission of radiation") is a source of light with a single color, meaning with a single wavelength and a single frequency. A laser also arranges all the light waves it produces to be "in phase", so that all the wave tops coincide and reinforce each other, giving very intense light.

lens of eye 3, 5.1.3: The lens in the eye helps focusing the incoming light onto the retina to achieve sharp vision. Muscles unconsciously adjust the shape of the lens for that purpose. When the lens does not have the necessary flexibility, corrective glasses can be worn as compensation. (See also **eye**.)

light 2.1: Light is electromagnetic radiation in the visible part of the electromagnetic spectrum. It is therefore a kind of wave that mixes electric and magnetic components. (See also **electromagnetic radiation**.)

light sources 2.5, 6.3: Light is created by various physical or chemical processes, such as: nuclear reactions in the Sun, chemical explosions, fire, lasers, lightning, electroluminescence in electronic displays, *etc.*

luminance 2.2, 4.1, 6.2.3: See **brightness**.

M cones 8.2: See **cones**.

mental model 10.3: The 2D physical image observed by the eye's retina is processed by the brain into a 3D mental model of the observed scene: we see that model as a mental image. (See also **vision** and **optical illusions**.)

metamerism 2.5: When two different mixes of colors produce the same apparent color, they are called metameric. An example of metamerism is pure yellow compared with a mix of red and green:

both appear yellow because both excite our eyes' red and green cones.

moiré patterns 3, 9, 10.1.1: Moiré patterns are wavy patterns that are commonly seen in sheer curtains and other situations where meshes overlap each other, including on television screens and computer displays. These wavy patterns are very mobile when the meshes move even slightly. (The French word moiré means wavy.)

neural networks 3, 5.1.5: Nerves (which consist of cells called neurons) form extensive networks across the body and especially in the brain. These neural networks operate in a way similar to computers, by repeatedly combining signals from various sources (such as from different parts of the eye's retina) into new signals. This information processing results in the mental images that we see in our minds.

number of colors 4.3, 6.2.2: It is claimed that humans can distinguish over 10 million different colors, if we include intensity levels; ignoring intensities, we may distinguish over 50,000 hues. The RGB color system commonly provides close to 17 million colors (including intensity levels), readily available for everyday word processing and graphics. (See also **colors**.)

optical illusions 10: There are many types of optical illusions: a common feature is that the mental image that we see in our minds includes contradictions with the original scene being observed. So what we think we see is surprising, and therefore often entertaining.

paint color 6.1: Colored paints are usually mixed from a limited set of paints with basic colors. The CMY and RGB color schemes can be used to characterize paint colors more precisely. (See also **colors** and **ink color**.)

pattern 10.1.1, 10.3.4, 10.3.5: See **structural pattern** or **pattern recognition**.

pattern recognition (in optical illusions) 10.1.1: An important step for the brain to interpret a 2D image in terms of a 3D model is to recognize in the image some familiar structural patterns that can be used to mentally build the model. For example, a 2D image may suggest a dining room composed of a table and chairs: these structures are patterns used to construct a model of a 3D kitchen.

In the Penrose triangle the 2D image suggests straight bars to be assembled in 3D space, but it turns out that these bars cannot be put together without violating known geometry, thus producing an optical illusion. (See also **vision** and **optical illusions**.)

perceived image 3, 10: The perceived image is the mental image produced by the brain based on the observed scene: the brain interprets the physical 2D image and constructs a mental 3D model of the observed scene; the result may be substantially different from the original scene. (See also **vision** and **optical illusions**.)

photography 5.2, 5.3: Photography makes a record of a scene in the form of an image on paper or film (in film and movie photography) or in electronic form (in digital photography). (See also **recording images**.)

pixels (picture elements) 3, 5.3, 6.2.1: A digital camera breaks up a scene into detector pixels (pixel is short-hand for picture element). The pixels are arranged in a regular grid of dots, each dot having a single color. An electronic display similarly arranges the image on a regular grid of emitter pixels. A digital picture stored in a graphical file may use yet another grid of image pixels. The detector grid, image grid and emitter grid are usually different from each other, so conversions (performed by a computer's graphics card) are needed between them to display a recorded image.

pupil 2.4, 8.3: See **eye**.

pure colors 2.5: Pure colors in this book refer to colors with a single frequency (namely, a single wavelength): the pure colors are the hues that are separated into a solar spectrum by a prism. (See also **colors**.)

radiance 2.2: See **brightness**.

rainbows 2.5: A rainbow is similar to a solar spectrum, since sunlight is split up into colors by raindrops. However, in rainbows the colors are slightly smeared out due to the spherical shape of the raindrops instead of a prism, so the colors of the rainbow are slightly different. (See also **solar spectrum**.)

recording images 3, 5: Eyes record images by means of molecular detectors called cones and rods in the retina. Film cameras use chemical reactions to produce an image on film or paper after development. Digital cameras produce an image by means of tiny electronic detectors. (See also **detecting light and color** and **electronic recording**.)

reflected colors 2.4, 2.6, 4.2: Light reflected by an object is often changed in color by that object. The light reflected when an object is illuminated by white light is usually considered to be the object's own color; but illumination with another color gives such an object a different color. During reflection, an object absorbs some hues from the incoming light: the reflected light then has a color given by the remaining hue mix. (See also **colors**.)

resolving power of eye (visual acuity) 5.1.2: The eye's resolving power (also called visual acuity) is the sharpness of the image produced on the eye's retina. Specifically, it describes how small are the details that you can distinguish in a scene. (See also **eye**.)

retina 3, 5.1.2: See **eye**.

RGB color system 4.1, 6.2.3: The RGB color system combines the three colors red (R), green (G) and blue (B) to compose many visible colors. It is used primarily for electronic displays. (See also **colors**, **CMY color system**, and **HSL color system**.)

rods 2.4, 3, 5.1.1: The rods are similar to the cones in the eye's retina: they are molecular detectors of light, but the rods have little sensitivity to color and more sensitivity to weak light of any color. They are mainly used at night and in dark environments. They give a gray image. (Their name comes from their rod-like shape.) (See also **cones**.)

S cones 8.2: See **cones**.

saturation 4.1, 6.2.3: The saturation of a color describes how colorful it is. A color is saturated when it contains no gray: it is then most vividly colorful and has the least "fog". A color has low saturation when it contains much gray, as if seen through fog. (See also **colors**, **hue**, and **brightness**.)

secondary rainbow 2.5: Rainbows often exhibit a secondary rainbow beside the more familiar primary rainbow. The secondary rainbow is due to sunlight that reflects twice inside raindrops before exiting, instead of once as in the more intense primary rainbow. (See also **rainbows**.)

sensitivity of cones and rods 6.2.1, 8.2: The cones and rods in the eye's retina have different sensitivity to different colors. The "red" cones (also called L cones) are most sensitive to red light, and less sensitive to green and blue light; likewise for the "green" or M cones, and for the "blue" or S cones. In addition, the "green" or M cones are about 2 times as sensitive to light as the "red" or L cones, and about 6 times as sensitive to light as the "blue" or S cones. Rods are more sensitive to weak light of any color (which is useful for night vision), and less sensitive to differences in color. (See also **cones**, **rods**, and **colors**.)

shades 2.3: Shades of colors refer to small variations in brightness, in contrast to changes in hue, called tints. (See also **tints**.)

sharpness of vision 5.1.2, 5.1.4: See **resolving power of eye**.

sky colors 2.8: The sky is illuminated mainly by the Sun. Air scatters sunlight in all directions. Blue light is scattered more than green light, which is scattered more than red light, explaining the color of the sky depending on how far through air the sunlight has traveled. (See also **Sun color**.)

solar spectrum 2.5, 6.2.2, 7.1: The Sun emits light of all visible colors, from red through blue. A prism splits those colors (hues) and spreads them out in the form of the solar spectrum. Each hue (red, orange, yellow, *etc.*) has its own wavelength or frequency. The solar spectrum is one (small) part of the electromagnetic spectrum, which also includes infrared and ultraviolet light, among others. (See also **colors** and **electromagnetic spectrum**.)

sources of light 2.5, 6.3: Light is created by various physical or chemical processes, such as: nuclear reactions in the Sun, chemical explosions, fire, lasers, lightning, electroluminescence in electronic displays, *etc.*

spectrum 2.5, 7.1: A spectrum is a spread of light (or other types of waves) according to wavelength or frequency, as in the solar spectrum, which is a part of the electromagnetic spectrum. The spreading of light can be performed by a glass prism. (See also **electromagnetic spectrum** and **solar spectrum**.)

specular reflection 4.2: Reflection of light is specular when it is mirror-like: light is reflected in a well-defined direction (at an exit angle equal to its incoming angle). Specular reflection contrasts with diffuse reflection, in which light is reflected in many directions. (See also **diffuse reflection**.)

stereoscopic vision 3, 5.1.4, 10.3.1: With two eyes, the brain receives two slightly different views of the same scene. This stereoscopic vision gives the brain useful depth information to reconstruct the 3D scene in our mind. A single eye does not provide such 3D information. (See also **vision** and **depth information**.)

stroboscopic effect (optical illusion) 10.1: If the lighting of a scene is turned on briefly and repeatedly (which is called stroboscopic lighting), we get a sequence of snapshots; this is similar to a movie, which is also a sequence of snapshots. If at the same time repetitive motion occurs in the scene, such as the rotation of a wheel, the motion under stroboscopic lighting may appear very different from the real motion. As an example of the stroboscopic effect (frequently seen in movies), a wheel may appear to turn slowly, backward or not at all: this is the wagon-wheel effect. (See also **optical illusions**.)

structural pattern 10.1.1, 10.3.4, 10.3.5: The brain analyzes every 2D image sent to it by the eyes: one important function of the brain is to extract familiar features, such as 3D structural patterns, to build a mental image of the scene from familiar components. For example, the brain will try to identify people, buildings and trees made of simpler structural patterns like faces, walls and branches. However, the brain may make mistakes in assembling a 3D mental model, creating an optical illusion. (See also **optical illusions**.)

subtractive process (CMY) 4.1: The CMY color system based on cyan, magenta and blue primary colors subtracts these three colors from white light. This contrasts with the additive process used in the RGB color system, which adds red, green and blue to black. (See also **additive process (RGB)**.)

Sun color 2.8: Sunlight is white before it enters the Earth's atmosphere, and thus contains all hues of the solar spectrum. The air scatters sunlight in all directions. Blue light is scattered more than green light, which is scattered more than red light: therefore, the Sun looks slightly yellow high in the sky (because some blue has been removed from the white light), or orange/red near the horizon (because much blue and green have been removed from the white light). (See also **sky color.**)

sunlight 6.3: Sunlight is electromagnetic radiation, which is a kind of wave with electric and magnetic components. Sunlight falls in the visible part of the electromagnetic spectrum. (See also **electromagnetic radiation, electromagnetic spectrum** and **solar spectrum.**)

tints 2.3: Tints of colors refer to changes in hue, in contrast to variations in brightness, called shades. (See also **shades.**)

ultraviolet light (UV) 2.5, 3, 5.1.4, 7.1, 7.3, 7.4: The solar spectrum extends beyond the range of visible light, which spans from red (at lower frequencies) to blue (at higher frequencies). Frequencies above blue include invisible ultraviolet light, x-rays, *etc.* (Frequencies below red include invisible infrared light, radio waves, *etc.*) (See also **electromagnetic spectrum, waves** and **infrared light.**)

visible colors 2.5, 7: For humans, the visible range of colors normally stretches from red to blue. This is the range of electromagnetic radiation in which the Sun produces the strongest light. (See also **electromagnetic radiation** and **colors.**)

vision 5.1: Vision covers several functions, primarily: collecting light from a scene (by lenses, *etc.*); detecting the light's intensity (brightness) and color composition (by cones and rods in the eye's retina); and processing the resulting electrical signals into a mental image (by the brain).

visual acuity 5.1.2: See **resolving power of eye.**

wagon-wheel effect (optical illusion) 10.1: See **stroboscopic effect.**

wavelength (of light or wave) 2.1, 2.5: See **waves.**

waves 2.5, 4.2, 5.1.3, 6.2, 6.3, 7.1, 7.2, 7.3, 8.2, 8.3, 9: Light is composed of electromagnetic waves that travel at high speed, including through the vacuum of space between Sun and Earth. All waves have a wavelength that measures the distance between successive peaks or valleys in the wave. All waves also have a frequency, which measures how many peaks or valleys pass a fixed point in one second. Wavelength and frequency are inversely related: a wave with a longer wavelength has a lower frequency. Other kinds of waves include sound, water waves, and earthquake waves. A moiré pattern may often be viewed as a wave. (See also **electro-magnetic radiation** and **moiré patterns**.)

white 2.5: White is the color of sunlight before it enters the Earth's atmosphere. When split by a prism, sunlight is seen to consist of all solar spectral hues from red through blue. White can also be composed of the right amounts of red, green and blue basic colors, as in the RGB color system. There is no upper limit to the intensity of white (or of any hue). Weak white is gray. (See also **colors**.)

white balance 2.1, 5.3, 6.2.1, 8.3, 9.1: The apparent color of an object is changed by the color of the illuminating light (see also **reflected colors**): in particular, a "white" object will take on the color of the illuminating light. The brain and many digital cameras try to compensate for the color of illumination, so that white still appears to be white: this is white balancing.

x-ray imaging 7.1, 7.3: X-rays penetrate living matter relatively easily, casting a shadow on a film that contrasts denser matter *versus* lighter matter. X-ray images are usually inverted (black and white are exchanged), because we see small variations in dark light better than in bright light. (See also **x-rays**.)

x-rays 2.5, 7.1, 7.3: X-rays are a part of the electromagnetic spectrum that is well outside the visible range and beyond the ultraviolet part. X-rays are relatively very energetic and therefore very dangerous to living matter in heavy doses. (See also **electromagnetic spectrum** and **x-ray imaging**.)

1

Our Goals

Let's start our exploration of physics with colors: in daily life, colors are all around us and we experience them mostly as joyful; think of "colorful"!

I invite you to open your mind to the fascinating world of colors, which will lead us into the interesting realm of light and vision. The colors and light will also lead us to entertaining and puzzling optical illusions.

Think of those exciting days when you, maybe as a child, entered a new home or new playground and rushed around to uncover its mysteries, secrets and pleasures. Try to do the same here with colors: explore and enjoy the colors of nature!

You may jump around this book as you please: there is no need to read from beginning to end, and you may skip all details. I have highlighted the important questions and answers for easy guidance: you may use them to locate topics of interest to you. You will find more guidance in the Table of Contents, the Index, and the List of Concepts, Connections and Terminology. You may also start by looking for a topic that strikes your fancy in the following list of frequent questions concerning colors, light and vision (feel free to jump directly to the indicated Chapters and Sections):

- How are the color of the sky and the color of the Sun related? Find out how the red, green and blue components of sunlight behave differently in air (Section 2.8).
- What do 20/20 and 6/6 vision mean? We will learn what good vision really means and how it is quantified (Section 5.1.2).
- Do two people see the same object as having the same color? In general not, because their eyes' color detection may differ (Section 5.1.1).
- Likewise, a repainted car panel may appear to have the original color to one person, but may look different to another person (Section 5.1.1).
- Can we distinguish older from newer documents by their color? Indeed: inks and paints age by losing their "colorfulness", called saturation, and this also allows faking the age of documents (Section 6.2.3).
- What are the colors of a rainbow and how do they come about? We will show how rainbows relate to the solar spectrum (Section 2.5).
- What do optical illusions tell us about our eyes and brain? This entertaining subject will give us fascinating insights into how our brain and nervous system work (Chapter 10).
- Did you realize that your eyes normally can only read about one word at a time, rather than a whole sentence or line of text? Indeed, and we will find that the reason lies in the tiny "fovea" in the back of our eyes (Section 5.1.2).
- Did you know that the lens of your eye only does about 20% of the job of focusing an image on your eye's retina? The other 80% is actually done by the curved outer surface of your eye: the cornea (Section 5.1.3).
- You have seen pictures made with x-rays, infrared light and perhaps ultraviolet light, but how can such invisible light actually be made visible? We will describe how invisible light is "shifted" into the visible range (Chapter 7).

- Have you ever wondered what color-blind and color-deficient people see? We will discuss this important topic, as about 5% of people suffer these conditions (Chapter 8).
- Do you know what cataracts are and what people with cataracts see? We will discuss this also (Section 5.1.3).
- How many different colors are there? We will see that there are infinitely many colors, and we will show how you can easily produce almost 17 million colors on your computer by yourself (Sections 2, 4.3, 6.2.3).
- Are just three basic colors sufficient to compose millions of other colors? Yes, and we will learn how that is possible (Section 2.3).
- Why are just three basic colors sufficient to compose all the colors that our eyes see? We will connect the three basic colors to the three types of "cones" in our eyes, which detect the fundamental colors red, green and blue (Section 2.4).
- How do modern television sets produce more vivid colors and how can this enrich digital painting? We will explain how new technology has expanded the richness of colors, not only for ultra-high definition television viewing, but also for electronic painting (Section 6.2.2).
- What is white light? We will learn how white light is related to the Sun's light through the "solar spectrum" (Section 2.5).
- What is laser light? We will also connect laser light to the Sun's light (Section 2.5).
- Are the colors produced on computers, television sets and smartphones different from those produced by painters and printers? While they are visually very similar, these colors are actually produced quite differently (Sections 4.1, 6.1, 6.2).
- What are "complementary colors" used by artists and designers? We will see how contrasting colors are used to produce a more dramatic impact in paintings and other images (Section 2.7).
- Why does your inkjet printer require black ink (in some cases even two different black inks), when three color inks should be sufficient? This question relies on understanding how a few colors are combined to form many more colors, including even black, gray and white (Section 6.1).

- Is black really a color? We will view black in two ways: absence of light and color, as in electronic displays (computers, television sets, smartphones); or absorption of light and color, as in paint and printers (Section 2.5).
- What is gray? As we will discuss, gray is simply weak white (Section 2.5).
- What is the color of a surface that reflects light? We'll see that reflected light is a combination of the color of a surface (for example, paint) and the color of the source of the light (for example, orange light of the setting Sun or a lamp's colored light) (Section 4.2).
- How do we record color in an image? This fascinating topic differs between our eyes, old chemical photographic film and modern digital cameras, so we will have interesting discussions about this topic (Sections 5.1, 5.2, 5.3).
- What is the meaning of those mysterious picture file names ending in .JPG, .GIF, *etc.*, that accompany your digital photographs? We will explain how these endings actually save a lot of computer space on your smartphone and on your computer by compressing your pictures (Section 5.3).
- And what is the meaning of the number of megapixels (MP) that camera makers advertise? We will see that one pixel is one "picture element" of a digital image, that a megapixel is a million pixels and that the more such pixels your camera has, the more detail your pictures will show (Section 5.3).
- Why would it be smart to use white-on-black text on your smartphone, tablet or pad, instead of the usual black-on-white text? We will learn that this extends the battery life (Section 6.2).
- Do you know what animals see? We will find that there is great diversity among animals in this respect: for example, many familiar animals only see two basic colors, such as yellow and blue (and their combinations), while others see as many as a dozen basic colors (and their combinations) (Section 5.1.4).
- Are pictures taken with different cameras identical or different? In fact, they are quite different and there are many reasons for this remarkable variety, as we will discuss (Chapter 3).

- You certainly have seen so-called moiré patterns, for example the rapidly moving wavy patterns in sheer curtains that swing in a breeze, but have you ever wondered what they are due to? We will dive into this mysterious but pretty topic (Chapter 9).

To help you along, I will ask ***many questions, highlighted in this manner (bold italic)***: I highly recommend that you first try to answer them yourself, before reading my discussion following the questions, and before jumping to the **answers, highlighted in this manner (bold blue)**. By critically thinking yourself, you will understand the physics of colors, light and optical illusions much faster and better. So, take the time to ponder my questions: they may seem simple at first sight, but they contain important physics that is easy to understand, and which we will discuss after you have had a chance to think about them. The physics will then gradually emerge from our discussion. This approach of investigating by yourself will improve your understanding and your recollection of the interesting physical principles involved. **So: think along and enjoy discovery!**

Each Chapter is also concluded by a convenient summary Section called "What have we learned in this Chapter?".

Interestingly, much of science has developed historically exactly through such questioning. Over coffee or a meal, scientists often ask each other such questions; also, a scientist will ask himself or herself such questions. For example: How does this work? Does it always work this way? Can I change this? What if I changed that? Why does this happen? How come that does not happen? And so on! Scientists often ask straightforward questions like children do: science is a fun game of exploration. Besides entertainment, the reward is satisfying your curiosity and understanding daily observations around you. Also, you will better appreciate all the modern machines invented by man to exploit such understanding.

You will certainly have more questions as you read. To help you answer those questions, consult the List of Concepts, Connections and Terminology at the beginning of this book: it will give you useful definitions and relationships, and refer you to the sections where they are further discussed. More terms are contained in the Index at the end of this book. Furthermore, a list of general References and Resources (books, videos, websites) is provided at the end of this book, so you can extend your reading. In particular, animations,[1] cutouts (*see* footnote 1), and videos[2] pertaining specifically to this book are available online; the most important animations are included in these videos.

NOTE: Boxed text like this one adds some details, or a deeper discussion, or another point of view which may interest you; these boxes also give a more in-depth flavor of physics. If you wish, you may skip these boxes, as such information is usually a bit more technical or advanced.

WARNING: In this book about colors, it is helpful to see the illustrations on an electronic display: the colors will then be significantly more accurate and vivid than when printed on paper.

---

[1] Animations and cutouts are available to buyers of this book at https://worldscientific. com/worldscibooks/10.1142/12316#t=suppl (for more details, see References and Resources on page 257).

[2] Videos on Everyday Physics by Michel A. Van Hove at YouTube: https://www.youtube. com/playlist?list=PLWIjtByUJvcuf0GCjqGJHNNXMdT1NIsx8.

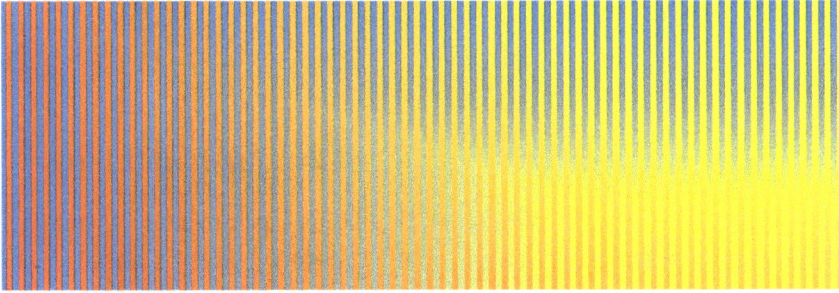

2

What is Light and What Colors Exist?

We start our exploration in this Chapter by asking what light is and what colors of light we know from our daily experience, such as red, purple, green, yellow and gray. Such colors are seen reflected from solid objects such as plants, furniture and clothes, in liquids such as water and oil, in the sky and clouds, and coming from light sources such as the Sun and lamps.

We will then try to group those colors in categories called hues, such as reds versus blues, and degrees of brightness (from dark to bright). We will show how sunlight can be decomposed into many hues, from red to green to blue: we call them "pure" colors; the pure colors can then be combined to form more complex visible colors, such as white, purple and brown.

We will discover that three colors are more basic than all the others: red, green and blue. The reason is that our eyes have three types of light detectors called "cones" which are most sensitive to red, green and blue, respectively. All other visible colors can be obtained by mixing red, green and blue; even black, gray and white, which we often call colorless, are made up of red, green and blue. Two videos are available online to illustrate these important points.[1]

— ·}} {{· —

2.1 How many colors can you distinguish and name?

Look at the following picture, Figure 2-1, which I took in a flower market. It is full of **colors**: try to name as many different colors as possible; do not limit yourself to this picture, but think of all flowers and all other objects. You may start with a few basic colors, such as: red, yellow, green, blue, black, gray, and white. But you surely know many more colors, such as: gold, silver, navy blue, salmon pink, and turquoise. Painters and decorators will be able to think of a very long list of color names!

Figure 2-1: Flowers in a shop.

Do you think that the color names are unique and universal? Think of colors defined for paints, ceramic tiles, automobiles, wallpaper, cloth and lipstick: do all manufacturers use the same color names (let's ignore different languages)? Actually, there exists no universal list of colors and color names based on precise and unique physical principles. If you search for color lists on the web, you will find a variety of such lists:

---

[1] See videos "How many colors can you tell apart?" and "Can we make all colors by using only red, green and blue?" on Everyday Physics by Michel A. Van Hove at YouTube: https://youtu.be/aNJBvxRs3T8 and https://youtu.be/nrsCWRCUESo.

some for paints, others for television and computer displays, *etc*. A color such as "salmon pink" is not defined in a unique way, so that the color "salmon pink" likely will vary from one paint manufacturer to another. In fact, no two salmon fish have identical pink colors! And different parts of the same salmon have different varieties of pink, so how could you define "salmon pink" precisely anyway? Even a basic color such as "blue" is not uniquely defined: there is a range of different colors "blue".

One reason for the absence of a unique list of colors is that we may perceive the same color in different ways under different circumstances. This may be due to different lighting conditions: the color of the same object seen under the Sun, under blue lamps or under green trees will often be very different. Another cause is the material on which colored light falls: blue light reflected from a pink screen will look very different than on a white screen; so, reflected color depends on the reflecting material.

More subtly, the color of paints also depends on the exact pigments used, while the colors recorded by a camera depend on the detector materials and software that the camera uses. Some readers will remember that photographic Kodachrome and Fuji films gave different colors: the first were "warmer" (more reddish) while the latter were "colder" (more bluish).

In modern digital cameras, software offers many options to the photographer for "improving" the recorded colors, while by default it already "optimizes" the colors and many other aspects; an example of **white balance** is shown in Figure 2-2. Such optimization is a compromise that is welcome for most photographs, but will not work satisfactorily in all cases. The camera's "optimization" can be turned off, but the photographer has to remember to turn it on again afterward for more normal lighting situations.

Humans also differ in their color perception; color deficiency (such as difficulty to distinguish red from green) and color blindness (such as absence of sensitivity to green) are obvious cases. We will discuss these issues in Chapter 8, but let's continue our exploration first: there is much more to discover!

Now that you have made a list of color names, *can you categorize the colors?* One category could be the red or reddish colors. *What other categories can you propose?* Are the categories mutually exclusive, or

Figure 2-2: I made this picture of a stream in a forest with my smartphone, using its automatic white balance. The plastic bag which I found there was pure white, but became bluish in this photograph. The sky was mostly overcast and gray, so the blue color of the plastic does not come from the sky. The camera software is supposed to compensate for color changes due to different kinds of lighting, as our brain does very well; however, digital cameras generally make cloudy or shaded scenes look bluer than in reality, including fog, haze and snow.

do you find that some colors should belong to two or more categories? For example, could or should turquoise, which is a mix of blue and green, be categorized both as bluish and as greenish?

Before we move on, we should ask the important and basic question: **What is light?** Of course, light is that bright "stuff" that enters our eyes (and cameras) and makes colorful images of the world around us. **But what is this "stuff" called light? Can you touch it?** Not really! **Can you feel it with your hand?** Yes, if it is intense, as in front of a fire or hot oven or electric radiator, even before your hand catches fire.

You probably know already that light is electromagnetic radiation. This is a kind of wave that mixes electric and magnetic components; this kind of wave also includes **radio waves**, **x-rays**, and more, which are invisible to our eyes. However, the wavy shape of electromagnetic waves cannot be seen, unlike waves on the surface of water. That is because we don't actually see these waves themselves; instead we only see their impact when they hit the retina in our eyes; nevertheless we call them waves of **visible light**, since our eyes record their impacts and sends that information to our brain.

How do electromagnetic waves work? Think of waves on the surface of water. If you place a small piece of wood on that water surface, it will bob up and down as waves pass it: the water waves push the piece of wood up and down, again and again; in addition, this bobbing piece of wood will create new smaller waves that radiate in all directions. An electromagnetic wave is similar: if that wave passes an electric charge, that charge will be pushed back and forth; a microscopic magnetic needle will also be twisted back and forth by that wave; the moving electric charge and magnetic needle will also send out new electromagnetic waves. Tiny electric charges and magnetic needles exist in many materials, in large numbers, so that electromagnetic waves are strongly affected by such materials. In addition, even in vacuum, the changing electric component creates a new magnetic component, and *vice versa*.

We can still ask: *What is the length of electromagnetic waves?* The size of the so-called visible waves is actually tiny: their wavelength (the distance between successive wave peaks and valleys) is somewhat less than a millionth of a meter; a millionth of a meter is a thousandth of a **millimeter** or one **micrometer** and is similarly small if expressed in feet, inches or **mils**. Again, we cannot see this wavy character of electromagnetic waves, while we can see the shape of waves on the surface of water. (Radio waves can have much longer wavelengths, even longer than a meter, while x-rays are much shorter than visible light waves.)

Next, we can ask: *What is the color of light?* Color is a property of light that our eyes normally can detect, such as whether the light is mainly red or mainly yellow or mainly blue. As we will gradually learn, the color of light relates to its wavelength: visible light with a long wavelength is red, with a medium wavelength it is green, and with a short wavelength it is blue; we will see that yellow is in between red and green and indeed has an intermediate wavelength. But yellow can also be made by mixing red and green, and thus by mixing waves with different wavelengths: we call this possibility to make a given color (here yellow) by different combinations of other colors **metamerism**, as discussed in Section 2.5. The possibility to mix colors is due to our "cones", the color detectors in our eyes which we will describe further in Sections 2.4 and 5.1.

Besides color, light also has other important properties: light travels through transparent materials (such as glass, water and air) as well as through vacuum (such as through empty space from the Sun and other stars to the Earth); light travels in any direction at an amazingly large speed (almost 300,000 kilometers per second, or almost 190,000 miles per second in vacuum); light also carries energy (sunlight provides most of the energy we use on Earth, including energy from fossil fuels).

2.2   Can colors be mixed to produce new colors?

Do painters mix colors to produce new colors? You probably know the answer: think of the variety of colors that painters prepare on their palette starting from a few basic colors that they buy in tubes!

Indeed, we can mix colors to produce other colors. A good example is magenta, a purplish color which we can make by mixing red and blue. In the middle of Figure 2-3, the red and blue colors shown at left and right are mixed evenly (50%:50%), giving a somewhat dark magenta color.

red   dark magenta   blue

Figure 2-3: Red (left), dark magenta (middle) and blue (right).

The color "magenta" is usually considered to be brighter than the "dark magenta" shown here. They are actually the same basic color. The basic common color is often called **hue**; a given hue can have different **intensities**, also called **brightness**, **luminance** or **radiance**, ranging from dark to light.

When mixing paints, it is not possible to increase the brightness of the components: the brightness is just the average of the two components' brightnesses. To achieve a brighter mix requires starting from brighter components, for example brighter red (such as pink) and/or brighter blue (such as whitish blue).

We can also produce a continuous range of colors by smoothly varying the mix ratio from 100% red and 0% blue all the way to 0% red and 100% blue. This adds various purplish reds and reddish blues, as seen in the continuous red-to-blue color bar of Figure 2-4.

| red | dark magenta | blue |

Figure 2-4: Smooth transition between red, dark magenta and blue.

How does mixing paint work? The color of paint is due to the chemicals within the paint. A painter can produce on a palette the red-to-blue bar shown above by mixing varying amounts of blue paint into red paint: by choosing the correct mix ratio, the painter will get the desired color between red and blue.

Does this mean that the painter has created a new chemical for each color? Nature does not have a continuous range of molecules that produce all the desired intermediate colors. Instead, the mixed paint is still composed of separate red paint and blue paint, but the two paints are split up into tiny parts that the eye cannot distinguish. This is like pouring milk into coffee: the mix is not a series of new chemicals with a continuous range of different colors from coffee to milk and different tastes from coffee to milk, but it is a variable mix of tiny drops of pure milk inside the pure liquid coffee. Chemists call this an emulsion of one liquid in another. In some cases, mixing will cause a chemical reaction that produces a completely new chemical and a new color, but that will only create a <u>single</u> chemical and a <u>single</u> color, not a <u>range</u> of colors.

2.3 The color triangle

What is the effect of adding a third color into the mix? As you may already know, by further adding green as a third basic color, we can produce many more colors, such as those shown in Figure 2-5 as a **color triangle** with pure **red**, pure **blue** and pure **green** corners. A painter could make such a triangle by mixing different amounts of red, green and blue pigments at each point in the triangle: equal amounts of red, green and blue in the center; more blue toward the top; more red toward the bottom left; and more green toward the bottom right.

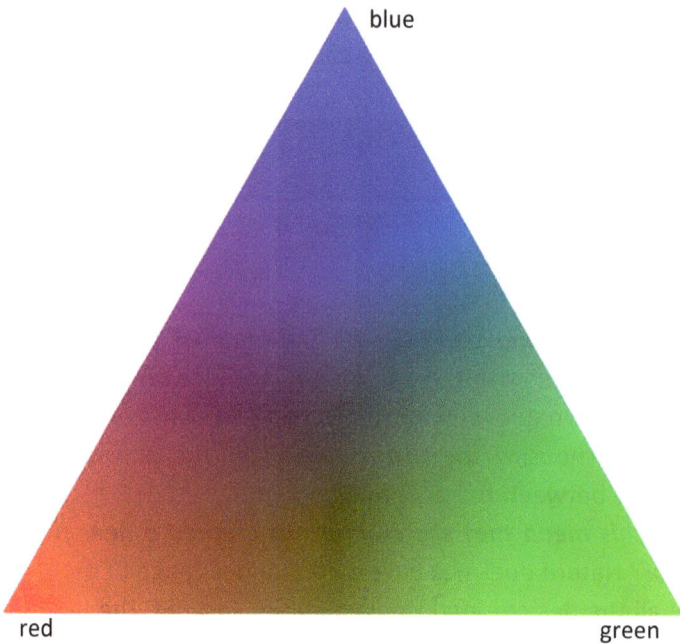

Figure 2-5: Color triangle based on mixing the pure primary colors red, green and blue. (*Source*: Courtesy of K.E. Hermann.)

Do you recognize the earlier red-to-blue color bar in the triangle of Figure 2-5? Consider the left edge of the triangle, from its red corner to its blue corner: the earlier red-to-blue colors are all lined up along that edge, including the purplish reds, the dark magentas and the reddish blues. But as you move to the right across the triangle, you see subtle color changes: gradually red and blue are removed while green is slowly added, until in the bottom right corner we have clear green with no hint of red or blue. Along the entire right edge we find no hint of red, and along the entire bottom edge no hint of blue, while the entire left edge has no hint of green.

Notice how rich in colors this triangular palette is. In the triangle, you now also see pairwise combinations of red with green along the bottom edge, and green with blue along the right edge, as well as many combinations of all three of these basic colors inside the triangle.

This triangular palette is indeed rich in colors, but: ***Do you think that this triangle includes all possible colors?*** Look at the collection of

Figure 2-6: A random collection of colored ovals.

many different colored ovals in Figure 2-6, and compare these with the color triangle of Figure 2-5.

Are some of these oval colors missing from the color triangle, Figure 2-5? If so, which kinds of color seem to be missing? Can you find color categories that are missing?

To provide some ideas, I have rearranged the colored ovals in Figure 2-7. The colored ovals are now arranged according to two principles: can you tell the two principles? Look first generally from top to bottom, and next from left to right.

In Figure 2-7, from top to bottom, we see a trend among the colored ovals: they go from bright to dark, from almost white to almost black. Such differences in brightness are often called **shades**. The colors near the top of Figure 2-7 are often called **off-white**, as they are very close to white, but with a hint of red or green, *etc.* Just below the off-white colors are **pastel** colors, which are still rich in white, but more reddish or greenish or bluish, for example. Near the bottom of this figure we find very dark colors, such as reddish black, greenish black and bluish black.

As you scan from left to right in Figure 2-7, you see colors gradually change from red to **yellow** to green to **cyan** to blue to **magenta** and back

Figure 2-7: An ordered collection of colored ovals.

to red. These color categories are called **hues**; the word **tint** is often used to express smaller variations in hue, as opposed to brightness. Notice how this sequence of hues follows the outer edge of the color triangle (Figure 2-5) counterclockwise from red to green to blue and back to red.

In Figure 2-7, I have split off the gray ovals to the right, because the grays cannot be found along the edge of the color triangle. However, as surprising as this may seem, grays actually do fit inside the color triangle: in Figure 2-5 the exact center of the color triangle is in fact one particular shade (intensity) of gray! This is indeed hard to see, but try this: if you curve your index finger to form a small opening and look at the color triangle through that opening, you can find the gray area about one third of the way up along the vertical center line.

Now we come back to the last question: *Are some of the ovals' colors (in Figure 2-7) missing from the color triangle (in Figure 2-5)?* Or, put differently: *Are all the ovals' colors also present in the color triangle, or does the color triangle miss some of the ovals' colors?* Think of the dark *versus* light colors: do you agree that the colors in the triangle are neither very dark nor bright? In fact, that color triangle has a limited range of brightness: it excludes very dark colors (very dark red, very dark green, black, *etc.*) and it excludes bright colors (yellow, light blue, pink, cream, rose, white, *etc.*). These very dark and bright colors are, however, present in the collection of colored ovals. Is brightness perhaps important, and not only hue?

2.4 Color: Hue *versus* intensity

What does all this mean? Do you perhaps see a role for the intensity of the color? To illustrate that intensity indeed plays a role, let's consider Figure 2-8. Here, we again have the earlier color triangle (Figure 2-5). However, I have doubled the color intensity within its central triangle. Thus, the color doubles its intensity upon entering from the outer triangle into the inner triangle. Suddenly we see many brighter colors that were missed before. In particular, near the midpoint between red and blue

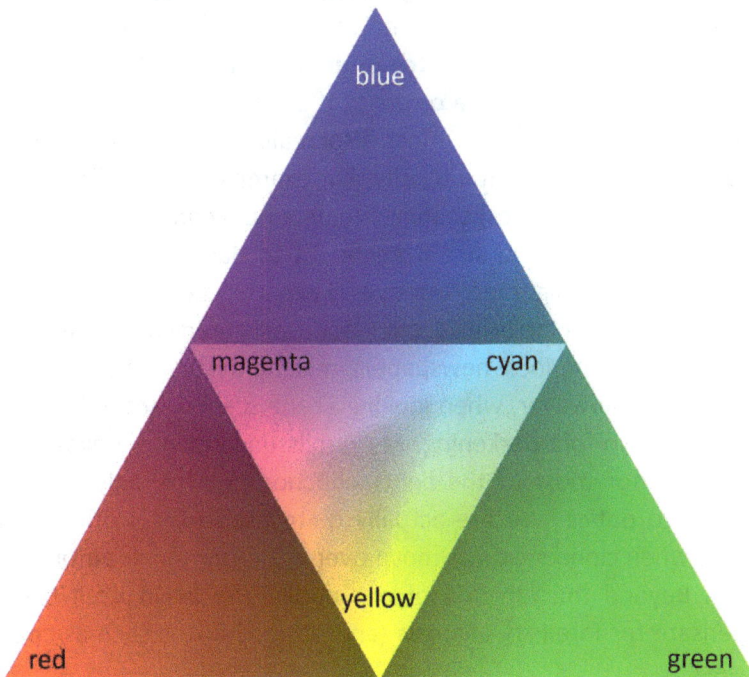

Figure 2-8: Color triangle based on the primary colors red, green and blue, as in Figure 2-5, but with doubled intensity in the center triangle giving cyan, magenta and yellow, as well as an infinity of their combinations. (*Source*: Courtesy of K.E. Hermann.)

we see "magenta": remember that we saw "dark magenta" earlier; increasing the intensity of "dark magenta" gives normal "magenta". We also see "yellow" near the midpoint between red and green, and "cyan" near the midpoint between green and blue. These three colors, cyan, magenta and yellow, as well as the basic colors red, green and blue, will turn out to be important later on: they form the basis of describing and making colors in inks, paints, computer displays, *etc.*

Incidentally, the midpoint of the brighter triangle in Figure 2-8 is again gray: here it is a brighter gray than in Figure 2-5.

Now, despite adding the cyan/magenta/yellow triangle, we are still missing some colors in these color triangles, namely the very dark and very bright colors: those that are almost black or almost white, including the extremes black and white. But we now have a strong hint telling us how to produce such very dark and very bright colors: by simply decreasing or increasing their intensity.

Indeed, the intensity or brightness of colors allows a much richer range of colors. In fact, we will find that the range of hues within the red/green/blue color triangle, together with their brightness, can well describe and produce all the colors that the human eye can detect.

On the other hand, we must also realize that our eyes are very tolerant of changes in light intensity. For example, our **pupils** (the holes letting light enter our eyes) change size depending on the light intensity: the pupils can double in size from bright to dark situations, allowing four times as much light into our eyes. I experienced this almost unconsciously during a solar eclipse: the moon obscured about 86% of the Sun (as I read in the local newspaper), making the sunlight almost eight times weaker. However, when looking at the scenery around me, it felt more like a two-fold darkening. My pupils unconsciously expanded to compensate for most of the light reduction, leaving only a two-fold reduction to notice. We are actually quite used to this effect: it also happens when cloud shadows move over us, as our pupils automatically expand. Beyond the function of our pupils, the **brain** itself can also compensate for intensity changes (and other changes such as contrast changes).

Are three basic colors, like red, green and blue, together with intensity, sufficient to define colors seen by the human eye? You probably already know why three colors are sufficient. The reason is the presence of three types of color detectors in our eyes: one type is sensitive mostly to red, another mostly to green and a third mostly to blue. These are the famous three types of **cones** in the retina at the back of our eyes. These cones correspond to what we perceive as red, green and blue, which are therefore called the **primary colors**. In addition, our eyes have **rods** that are more sensitive to the intensity of weak light and much less sensitive to the color of that weak light. The rods are specialized for night vision, but they make us see weak light as colorless gray.

So far, we have spent much time discussing various colors and their relationships, but you may be asking yourself: *Where does color actually come from?* Most of the colors that we see are those of objects around us: a tomato is bright red, a tree has various shades of brown and green, while a flower may be blue or yellow, for example. However, if we turn off all the light sources, including the Sun and the Moon (by hiding them), all those objects are black and thus invisible. This tells us that such objects only have color if they are illuminated. Once we realize this, we can answer: An object's color is the color of the light that is reflected by that object. Which color is reflected depends on the chemical composition of the object.

Of course, this is an incomplete answer, as it begs the further question: *Where does illumination come from?* We will discuss these questions in more detail in Sections 2.6, 4.2 and 6.3.

2.5   Black, white and gray; the solar spectrum

Can you now propose what the color black is? We have seen black in the colored ovals of Figure 2-7: when the ovals are arranged from bright to dark, black arises at the dark end of any color, including the dark limit of gray. The black ovals have no red, green or blue: in fact black has zero color of any sort. In a closed room without lights, you see nothing but blackness, because there is no light of any color. We can thus properly say that black is the absence of visible light.

Nevertheless, in printing and painting, we definitely need to apply what we call black ink or black paint. The physical function of black ink and black paint is to remove the reflection of light of all colors, leaving no light to reach the eyes. This removal is normally done by **absorption**, meaning that any light falling on the ink or paint will be converted to something else, primarily invisible heat in the ink or paint: a black surface absorbs light of all colors and therefore reflects no light of any color. We will return to this point shortly.

Do you know what white is? The color triangles of Figure 2-5 and Figure 2-8 do not show white, but some of the colored ovals in Figure 2-7 are so bright that they look very close to white.

Try to imagine where white light comes from. White paper and white paint only serve to reflect light that comes from another source: without light, both white paper and white paint look black, while when illuminated by red light they look red! So white paper and white paint do not by themselves tell us what is white.

We must therefore look further for the nature of white light. First: *Where does white light come from?* In daily life, white light comes from sources like the Sun and certain types of lamps. Second, we have an intuitive feeling for when light is white: white light is neither red nor yellow nor blue nor any other "color", so we feel that white is "colorless". Nevertheless, for convenience, we do talk of "white light" and "white color", as if white were just another color.

So we must dig deeper to understand white color and white light. You probably have seen the effect of a prism (a piece of glass with angled faces) on sunlight, similar to the photograph in Figure 2-9: a prism can split light into rays of different colors. This suggests that white light is a mixture of multiple colors; white paper and paint would then simply reflect all the multiple colors falling on their surface. Let's analyze this idea further.

Scientists call such a display a spectrum, for example the solar spectrum: a spectrum decomposes light into its hues. The brightness of each hue will also be visible in such a spectrum, and is very important for identifying the source of the light: the spectrum will tell whether the light comes from the Sun (which gives a slightly yellow/orange color due to scattering in the atmosphere), an old incandescent lamp ("warm" and slightly reddish), a mercury lamp ("cool" white with a bluish-green tint), or a sodium lamp (yellow as in street lamps).

For comparison with the spectrum of Figure 2-9, I show in Figure 2-10 the "spectrum" of a red **laser** (like the lasers used at check-out counters in shops or those used as pointers in classrooms): there

Figure 2-9: Photograph of light (arriving horizontally from top left) being split by a prism into rays of different colors, thus forming a spectrum of colors. The light comes from a mercury-vapor lamp: it contains more bluish-green and less red than pure white sunlight. (*Source*: D-Kuru/Wikimedia Commons under License CC-BY-SA-3.0-AT.)

is one spot of light, not a spreading of different colors, which would spread the laser beam into a long streak like the solar spectrum; thus there is no real spectrum, because there is only one color, namely one red. Indeed, a very important property of lasers is that they produce only a single color, such as red, unlike most lamps that produce white light consisting of multiple colors.

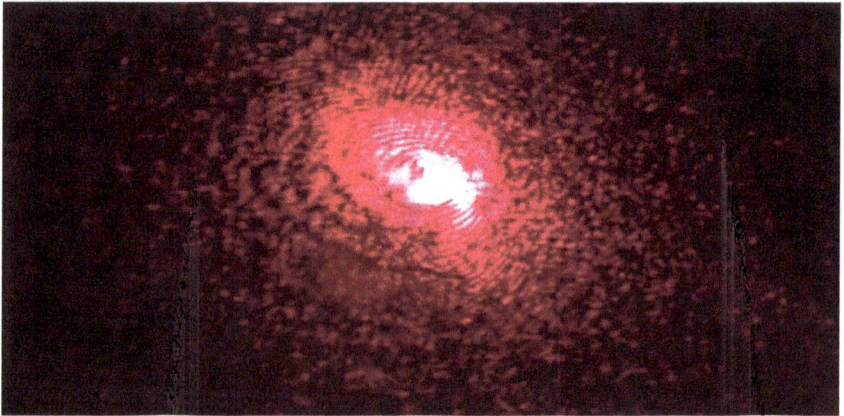

Figure 2-10: Photograph of a laser "spectrum" after bending through a prism, made analogously to Figure 2-9: only a single red color is present, with a single wavelength (about 650 nanometer in this case; the central white is due to camera overload and is thus artificial). Laser beams have a characteristic "speckle" pattern, due to the wave nature of light and the single wavelength of laser light: laser waves cause complicated ripples, similar to waves moving through each other on the surface of water.

A MORE TECHNICAL NOTE: You may object: *What is that white light in the center of the laser spot in Figure 2-10?* If it is white, surely it must contain at least red and green and blue, which should spread out into a colorful spectrum like that of the Sun! What is going on here? The answer lies in the camera: the laser spot is so intense that it triggers not only the red sensors but also the nearby green and the blue sensors, through flaring, as we will see in Section 5.1 in connection with lenses; see Figure 5-7). Furthermore, all sensors near the laser spot are overloaded (saturated) and give out a signal saying that they perceive the maximum allowed amount of red, green or blue light, respectively: those overloaded and therefore incorrect signals imply white light, as seen in the photograph. The same is true of eyes: our red, green and blue eye cones would be overloaded by laser light and — if not yet blinded by the intense light — send signals that the brain interprets as white light.

We recognize in the spectrum of Figure 2-9 some colors from the triangles and ovals above: from the left we see in particular the hues red, orange, yellow, green, cyan and violet.

Figure 2-11: Simulation of the visible solar spectrum, made with Microsoft PowerPoint.

I produced the spectrum in Figure 2-11 to better simulate professional solar spectra such as that of Figure 2-12. You can find in these spectra all the colors seen in Figure 2-9.

HERE IS MORE ADVANCED INFORMATION FOR THE CURIOUS READER: A spectrum spreads out the colors according to **wavelength** or **frequency** of **electromagnetic radiation** (we will describe electromagnetic radiation in another book). Indeed, light is a **wave** of electromagnetic radiation. A simple wave (for example a wave on the surface of water) has a wavelength which measures the distance between crests or valleys; a wave also moves constantly, such that at any given point it moves repeatedly with a regular period of time, which implies a repetition frequency (the frequency indicates how often the wave repeats per second). We will discuss waves in another book, as they are very important for vision, sound, telecommunications, medicine, electronics, *etc.*

As you go from left to right in Figure 2-11, at each point you see only a color that has a single wavelength or frequency in actual sunlight. (To each wavelength is associated a unique frequency, so wavelength and frequency are just alternative ways of describing the same thing.) Thus, hue corresponds to wavelength or frequency. Visible light is only a small portion of the **electromagnetic spectrum**: here we show and discuss only the visible portion. Light from the Sun extends farther to the left than shown: to the left of red, it continues into the **infrared (IR)**, **microwaves**, **radio waves**, *etc.*; sunlight also extends farther to the right of blue, into **ultraviolet (UV)**, **x-rays**, **gamma rays**, *etc.* We may loosely call these forms of light "**invisible colors**".

Interestingly, with electronic displays we cannot exactly reproduce pure colors that have a single wavelength or a single frequency: each material that produces light in electronic displays can only generate a range of multiple colors (hues): it thus emits a broader range of wavelengths or frequencies, meaning a mix of hues. By contrast, lasers can produce nearly pure colors of the solar spectrum, but only for some specific hues with very particular wavelengths or frequencies.

(Continued)

(*Continued*)

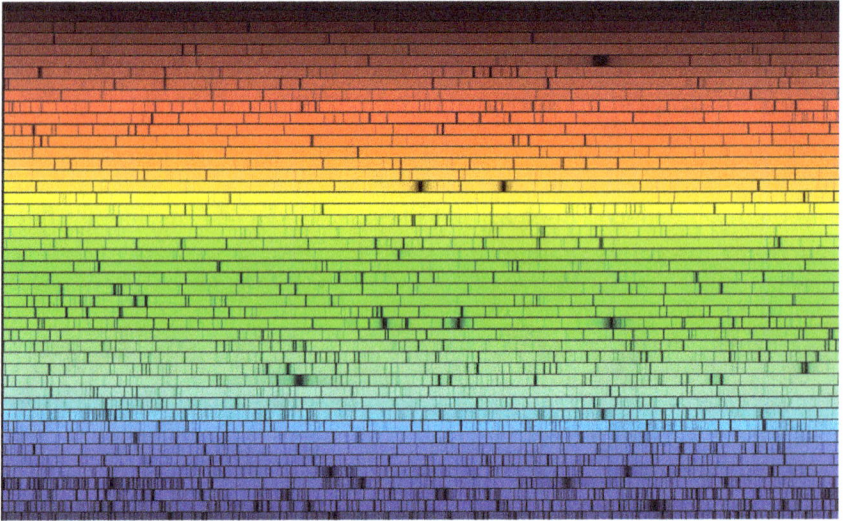

Figure 2-12: Photograph of the solar spectrum, by the US National Aeronautics and Space Administration (NASA). The spectrum, covering the wavelengths 400–700 nanometers, is cut up into 50 slices laid one below the other; each slice covers 6 nanometers. (This photograph is in the public domain; it is available with better detail/resolution at https://solarsystem.nasa.gov/resources/390/the-solar-spectrum/.)

Even the real solar spectrum is more complicated than a simple smooth distribution of colors. Indeed, the actual measured spectra of light coming from the Sun also have dark **absorption lines** (seen in Figure 2-12, which shows a professional and highly detailed solar spectrum from NASA): these narrow dark lines are due to nuclear reactions in the Sun and to gases or dust in the Earth's atmosphere: they show up at those wavelengths or frequencies where light is absorbed and does not reach us on the surface of the Earth. These absorption lines are not shown in my simulated spectrum of Figure 2-11; they are also not visible in my pictures above, because such absorption lines are too narrow to be detected with normal cameras or our eyes.

Are all colors present in the solar spectrum? Can you find the following colors in the spectra shown in Figures 2-9, 2-11 and 2-12: white, gold, silver, brown, pink, turquoise, or the color of skin? In fact, these and many other colors are not present, so we can conclude that the solar spectrum does not include all visible colors. We will therefore distinguish the "pure" colors, which are all the colors that exist in the solar spectrum, from the other colors, which are not found in the solar spectrum. We will find that the other colors are in fact mixtures of two or more "pure" colors: for example, "white = red + green + blue" and "brown = red + green". As we will see, this mixing reflects the roles of the three types of cones in our eyes, which are most sensitive to red, green or blue.

What do we mean when we say that the colors in the solar spectrum are "pure colors"? In a sense, the colors in the solar spectrum are in fact the purest possible. If you picked out a very narrow vertical slice of the solar spectrum in Figure 2-11, it would have one unique color (a single hue), with no admixture of any other color; and we could not compose that pure color by any combination of other colors from the solar spectrum. For example, we can select a very narrow slice of green near the center of the solar spectrum: that "green" would then be different from any other slice of the spectrum (including different from reddish-green or bluish-green slices) and it would not contain any color other than that particular green. In physics that unique color is given by a unique wavelength and a corresponding **frequency**: each color of the solar spectrum thus has a unique wavelength and frequency: scientists call such a color **monochromatic**, meaning single-colored.

We now must ask: *Where is the white color in the solar spectra of Figures 2-11 and 2-12?* In the spectra, the white sunlight has been decomposed into its parts, and those parts are not white. Thus, white is not an independent pure color: white does not exist as a single pure color in the solar spectrum. Instead, white is a combination of pure colors that are present in the solar spectrum. Similarly, brown and the many other colors do not exist as single pure colors in the solar spectrum, but are combinations of pure colors that are in the solar spectrum. For example, we may say that "brown = red + green": while brown does not exist as a pure color in the solar spectrum, it can be produced by mixing two pure colors, red and green.

Do we need to mix all the colors visible in the spectra above to produce white? If not all colors, then which colors do we need for making white? Clearly, the Sun mixes all colors from red to blue in order to form white, as we saw in the solar spectrum of Figure 2-12, which comes from decomposing all the colors present in sunlight. But are there other possibilities? Consider Figure 2-13. We see here three overlapping circles with the primary colors red, green and blue; where the circles overlap, they form the pairwise combinations cyan, magenta and yellow, already familiar to us. However, in the center where the three circles overlap, we see white! Indeed, if the red, green and blue have just the right intensities, they add up to what looks to us like pure white, even though in reality that white consists of only three distinct colors. So, three colors suffice to make white.

Can we also use cyan, magenta and yellow to make white? Look again at the circles in Figure 2-13. Notice how the three "petals" colored cyan, magenta and yellow overlap near the center, and together produce white there. We conclude that there are indeed more ways to produce white, such as mixing cyan with magenta and yellow. It is

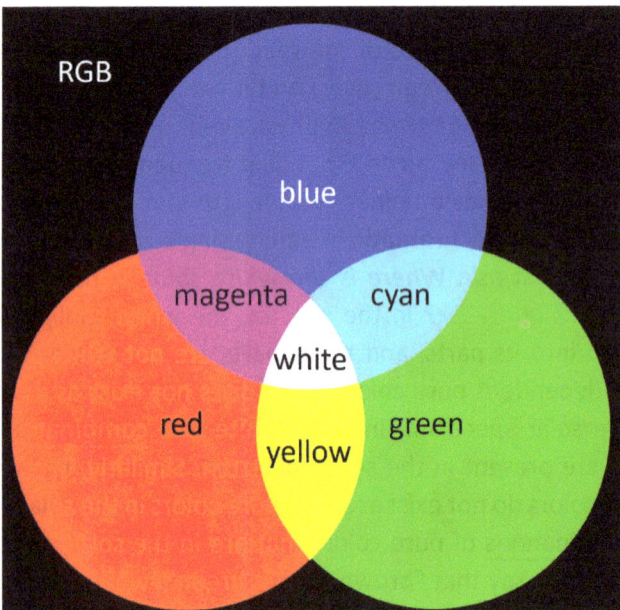

Figure 2-13: Overlapping primary red, green and blue circles, mixing to form cyan, magenta and yellow petals and a white center. This construction uses an additive system of red, green and blue colors, starting from a black background. We can imagine this image to be the result of three overlapping beams of red, green and blue light hitting a white screen: if the intensities of the three beams are just right, we get a white patch where they all overlap.

therefore not necessary to combine all colors of the solar spectrum to produce white. All we need is that the different colors that are mixed together contribute just the right amount of primary red, primary green and primary blue to excite our eyes' cones to the same extent as does natural sunlight. In particular, the right amount of just three colors can add up to pure white, as shown in Figure 2-13 with cyan, magenta and yellow.

Can you convince yourself that cyan, magenta and yellow also can mix to produce white? Remember that each of these is composed of equal halves of red, green or blue. Imagine that you are mixing paints in a pot. Any excess or deficiency of one of these colors gives a different color: for example, we get pink if there is a bit too much magenta, or very light blue if there is too much cyan.

It is of course also possible to make white with more than three colors, and not only with an infinite number of colors as in the solar spectrum itself. Furthermore, we can also make white with just two colors: adding the right amounts of blue and yellow will make white, as we see in Figure 2-13 where the blue circle overlaps the yellow petal (imagine the red and green circles replaced by just the yellow petal).

More generally, the possibility to produce the same apparent color (white in the examples above) by making different mixes of other colors is called **metamerism**. Another example is yellow itself, which is a pure color of the solar spectrum, but can also be made by mixing red and green: the reason pure yellow and a combination of red and green appear the same is that both excite our eyes' red and green cones, so our brain can't tell the difference between them.

We may conclude that white light does not have a pure color, but is composed of pure colors, such as red, green and blue from the solar spectrum; most white light from Sun and lamps is composed of many pure colors, in just the right amounts.

This is important knowledge for making white light, for example in lamps or in computer and television displays. We will look more into how white is made in Sections 4.1 and 6.2. For now, we ask another question.

Do you know what gray is? Think about it for a moment! Then look back at the colored ovals in Figure 2-7: to the right you see gray ovals lined up from bright (which is practically indistinguishable from white) to dark (practically indistinguishable from black). Could gray then be described as "weak black"?

If we take the point of view that black is the absence of light, then weak black is a contradiction: since black already has zero intensity, it cannot be made even weaker. On the other hand, if we think of black ink or paint, then too little ink or paint on a white background will produce gray instead of black, and we may then say that weak black simply means "not black enough".

Can we call gray "weak white"? That is realistic if we think of white ink or paint on a black background, the opposite of what we just discussed with black ink or paint on a white background: gray then means "not white enough".

But gray as weak white is also realistic if black is viewed as the absence of light, since white is intense light that we can make less intense. Let's think of a white page of paper in a totally dark room: that white page will look perfectly black. Now let a little bit of white sunlight enter the room and scatter from white walls, so you start seeing the page. The page will not look immediately white, but still quite dark gray. As you allow more sunlight into the room, the page gradually brightens up until it looks intensely white when placed directly in bright sunlight. The page's color has smoothly changed from black to white, passing through dark gray and then lighter intensities of gray. We can thus describe gray as low-intensity white. Figure 2-14 shows a color bar with all intensities of gray, from totally black to maximum white like a white page. (Can you tell where the color bar ends toward the right? What happens on a colored paper page?)

black   gray   white

Figure 2-14: A smooth progression from black to paper white.

Note that, similar to white, gray can be composed of three colors, such as three darker but equal intensities of red, green and blue. In fact, in both color triangles shown in Figures 2-5 and 2-8, the center points are gray; this is not at all obvious as you look at those triangles, but true.

Is gray colorless? We often think of gray as colorless, but is that really justified? Could we simply call gray a fourth primary color besides red, green and blue? After all, gray can be detected by a fourth type of detector in our retinas besides the cones: the "rods" (cones and rods are introduced in Section 2.4). But there is a significant difference: gray can also be detected as an equal combination of red, green and blue, meaning that light which excites our three types of cones equally is also interpreted by the brain as being gray (this is similar to yellow, which is a pure color of the solar spectrum, but can also be composed of primary red and green). Note that we might equally well declare white to be the fourth primary color, since gray is simply weak white. However, scientists do not normally declare gray or white to be the fourth primary color, since it behaves differently and can be composed of the three other primary colors.

So, gray (or white) comes from two different processes in the retina, one through cones and another through rods. The brain is clever enough to equate the two, so that we experience a smooth transition from the gray detected by the cones in bright light to the gray detected by the rods in dim light. A further difference is that our rods are not sensitive to bright light: otherwise our rods would "burn up" in bright light; this means that the rods filter out bright light, unlike our cones which are damaged by bright light.

Is brighter white possible? The right end of the black-to-white bar in Figure 2-14 merges smoothly into the white color of the page. This means that the brightest white we can make in Microsoft Word or PowerPoint is equal to the white of a paper page, whether on actual paper or on a computer display. Is it possible to create a brighter white than this? Remember that an actual paper page simply reflects all the light that falls on it. Therefore, if brighter white falls on that paper page, the reflected light will also be brighter, while retaining its white character, namely an equal mix of red, green and blue. You can try this yourself by focusing light through a lens (or some corrective glasses), or by shining a flashlight or your smartphone's light onto a white page: your white page will look brighter than without that additional light.

There is no upper limit to the brightness of light reflected from paper, at least until the paper burns because some of the light energy will be absorbed by the paper, heating it up.

Displays emitting light, such as a computer screen or television monitor, can be set to emit a brighter or darker white. On most monitors, you can adjust the intensity of the image: if you have not set it to maximum, it can be increased, making white whiter, but probably not by much because neither you nor the manufacturers want to blind your eyes! Other **light sources** can produce much more intense light, such as searchlights, spotlights, lasers (see Figure 2-15), welder's torches, explosions and the Sun. There is no limit to the brightness of white.

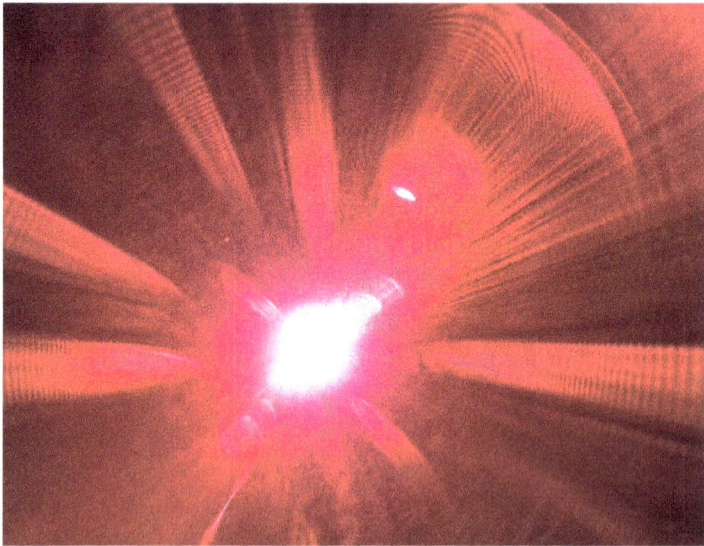

Figure 2-15: Low-power laser (emitting less than 5 milliwatts) shining straight into a smartphone camera (this is not recommended!). The laser light is purely red (with wavelength of about 650 nanometer; see the wavelength scale in Figure 8-2), but the camera records it as white in the center, showing overexposure of red, green and blue detectors, as explained earlier. The pretty reflections probably occur inside the camera; their fine ribbed structures are due to the wave character of light.

A consequence is that there is no way to define what the "proper" intensity of white must be: what looks white in a well-lit room may become gray when the lights are turned down; what appears white to you may be gray to someone else. Put differently: white is relative to your expectations or your prior experience. Someone living on snow will have a different concept of white from someone living in a jungle or in a deep cave.

Light sources can definitely be so strong as to blind you, essentially by destroying your eye's retina (think of sunlight and the need to use filters to view a solar eclipse without harming your eyes). And such sources can also produce very strong colors that are not white; a color filter is sufficient to remove unwanted colors from blinding white and still retain a very strong non-white color, while a laser already has a non-white color that can blind you. So, light of any color can be too bright for our eyes!

What are the colors of a rainbow? We have all enjoyed seeing wonderful **rainbows**, such as the splendid example displayed in Figure 2-16; notice the weaker **secondary rainbow** at top right and top left: the brighter main rainbow is called **primary rainbow**.

The colors of rainbows look very similar to those of the solar spectrum (see Figures 2-11 and 2-12), ranging from red through green to blue. The reason is that rainbows are generated in much the same way as a solar spectrum caused by a prism: raindrops act like prisms and therefore also spread out the colors of sunlight.

The primary rainbow results from sunlight that enters raindrops, then reflects once from the back of the raindrops, and exits the raindrops toward your eyes. The secondary rainbow is similar except that the sunlight reflects twice inside the raindrops before exiting, which changes the direction where it is seen; the second reflection also makes the secondary rainbow weaker, and reverses the color sequence — red is now inside and blue outside. The secondary rainbow is often too weak to see, while many people miss it because they concentrate on the primary rainbow: always look for a secondary rainbow outside the primary rainbow!

Rainbows are created not only in rain, but in any water spray, such as fountains, sprinklers and clouds, as long as the Sun shines on the drops. You can also create a rainbow with another source of light, such as a powerful lamp: then the rainbow will show the color composition of that other light source.

The spherical shape and multitude of raindrops complicates the rainbow's appearance slightly compared to the solar spectrum. The main effect is the smearing of each color in the direction from red to blue: red light spreads over the green and blue parts, while green light spreads over the blue part, so that the blue part is overlapped with red and green. Thus, green is more yellow, while blue is less saturated and therefore whiter or "foggier". Furthermore, this smearing of colors sends more sunlight inside the primary rainbow as well as outside the secondary rainbow, leaving a darker band between the two rainbows (this effect is relatively weak in the case of Figure 2-16). The faint thin pastel-colored lines that are sometimes seen inside the primary rainbow are caused by a wave effect of light: they are called **supernumerary rainbow**.

Figure 2-16: A double rainbow photographed in Alaska. The inner more intense rainbow is called the primary rainbow, while the outer, weaker rainbow is called the secondary rainbow; it is seen at upper left and upper right. Both rainbows can form full circles, when not obstructed by the ground. (*Source*: Eric Rolph, shared under CC BY-SA 2.5 at Wikipedia.[2])

Thus: Rainbows are a slightly smeared out version of the solar spectrum.

2.6   What is the color of a surface that reflects light?

We can now think about another question: *What is the color of a surface that reflects light?* Looking around us we see many objects with different colors. But we also see that those colors depend on the

---

[2] https://upload.wikimedia.org/wikipedia/commons/5/5c/Double-alaskan-rainbow.jpg.

lighting conditions: whether in sunlight or under a gray cloudy sky, in candlelight or neon light, *etc.* A white dinner plate will in fact look red in a red sunset and orange in candlelight! So, can we say that a surface has its own invariable color? After all, we do speak of a white dinner plate, a red ball and a blue toothbrush, as if those were unique and unchanging colors.

We have to accept the reality that an object's color in reflected light does depend on the color of the light falling on it. **When we speak of an object's color, we actually mean its color under white light,** rather than under any other kind of incoming light. That makes sense since we mostly observe objects in white light or nearly white light. We will discuss the color of objects in more detail in Section 4.2, after we have learned more about the relationship between colors. And we will look at different sources of light in Section 6.3.

2.7 Complementary colors

Have you noticed that certain color pairs contrast more than others? For instance, a red Sun stands out sharply against a blue sky. Often, workers wear yellow clothing to be more visible against common green, blue and gray surroundings. Rescuers also wear yellow, while red coloring is used to draw attention to firefighters and safety equipment.

Based on such observations, a list of contrasting colors has emerged: they are called **complementary colors**. Examples include: red *versus* cyan; blue *versus* yellow; and green *versus* magenta. Such pairs of complementary colors are somewhat subjective (for example, you may prefer to call red *versus* green a complementary color pair) and they depend to some extent on the brightness of the colors. We can make such color pairs more objective and independent of brightness as follows.

Take a look at the color triangles in Figure 2-5 and Figure 2-8, as well as the color circles in Figure 2-13. When you try to locate the complementary color pairs, you will notice that they tend to occur opposite each other. This can be better illustrated in the left-hand color circle of Figure 2-17: here we see the six basic colors arranged in a circle, in the same sequence as around the color triangles, from red *via* yellow, green, cyan, blue, magenta, and back to red. Now the above-mentioned color pairs are again opposite each other: red opposite cyan, *etc.* So we can call as complementary those color pairs that are opposite each other in the color circle (in Section 4.2 we will make this definition even more precise).

Color circle: complementary colors

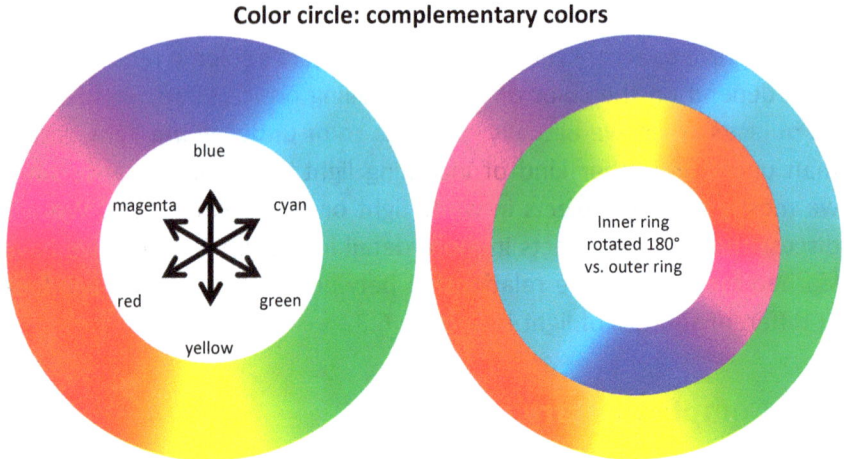

Figure 2-17: At left, a color circle shows complementary colors on opposite sides of the circle, such as red *versus* cyan. At right, the same color circle is overlaid by a smaller copy rotated 180 degrees, so that complementary colors touch each other along their circular boundary.

To better show the contrast between complementary colors, the right-hand image of Figure 2-17 takes the same color circle (outer circle) and copies it at the center after rotation by 180 degrees, so it is upside down and the opposite colors have been brought right against each other: everywhere along their mutual boundary we see a strong difference in colors. An important reason for this contrast is that a primary color present in one color is totally absent from the complementary color; for example, red is totally absent from its complementary color cyan. The explanation is: complementary colors use different sets of cones, in this case red cones for red light, but green and blue cones for cyan light.

Incidentally, in the color circle we can also recognize the solar spectrum (see Figure 2-11) wrapped around the circle, but with an extra magenta piece inserted to complete the circle.

Painters often utilize complementary colors, for example to empha-size contrast or draw attention to certain parts of a painting. Since our eyes constantly and rapidly move, complementary colors create rapid contrasts in different sets of cones. Complementary colors thus tend to excite, dramatize or create tension.

We will encounter complementary colors again when we discuss optical illusions in Section 10.1.1.

2.8 Sun and sky colors

We are familiar with the changes in the **Sun**'s and the **sky**'s colors as the day progresses. The daytime yellowish Sun and the bright blue sky are an example of complementary colors. On the other hand, at sunset (and sunrise), the Sun and the sky are both often red. *So how do the colors of the Sun and of the sky relate to each other?*

Figure 2-18 gives six schematic views of the Sun in the sky. The simplest situation is shown in the left image (a): we are viewing the Sun from outer space, such as from a satellite or space station orbiting the Earth. In this situation, there is essentially no air between the viewer and the Sun (we may also ignore dust, the Moon, planets and stars). We see the Sun as pure white, while the background space is pure black.

The air of the atmosphere is the main cause of changes in color of the Sun and sky on Earth. Air scatters light, meaning that light rays are bent toward different directions by tiny molecules and dust in the air. That is why sky, clouds and smoke are visible when illuminated by the Sun or any other source of light: they scatter the incoming light in all directions.

Now let's think of the **sunlight** as composed of red, green and blue components, remembering that white light consists of equal amounts of red, green and blue light: Figure 2-18 shows these three colors as red, green and blue lines and arrows of decreasing thickness as the light weakens and progresses from situation (a) to situation (f). Due to properties of electromagnetic radiation, air bends blue light more easily than green light, which is bent more than red light. So, when white light enters air, its blue component is bent away first (shown as a blue arrow in situation (b)): this gives the blue color of the sky in situation (b) of Figure 2-18. The remaining white sunlight has lost some of its blue component, leaving the complementary color yellow: this makes the Sun look slightly yellow. This situation is shown in parts (b), (c) and (d)

Sun and sky colors

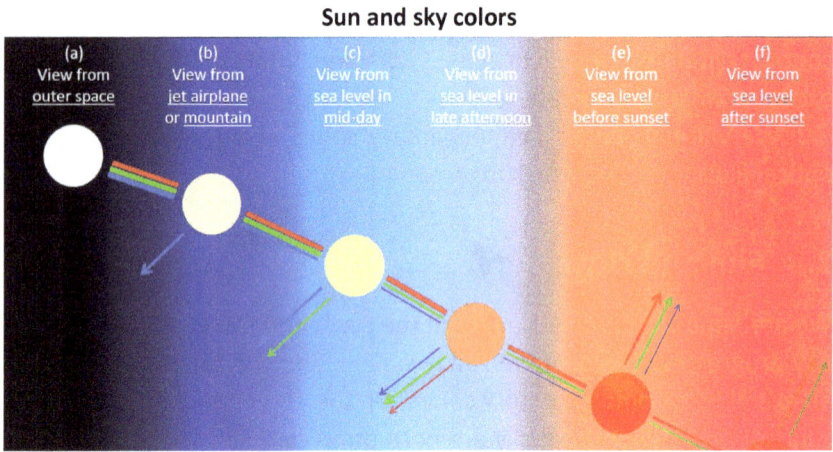

Figure 2-18: The six circles show the Sun in different situations in the sky, further described in the text. (Colors at sunrise are the same as at sunset, in reverse order, assuming the same weather.)

of Figure 2-18: these are typical views from high altitude in a jet airplane (or on a mountain), from sea level at mid-day and in late afternoon.

The closest most of us get to outer space is by flying in a jet airplane at high altitudes (typically 30,000–40,000 feet or 9,000–12,000 meters above the Earth's surface): there the air density has decreased to about a third to a fifth of what it is at sea level, so that we are then above most of the air of the atmosphere. At such altitudes, the air scatters a little blue light from the Sun, giving a deep blue sky (see Figure 2-19). This leaves the Sun looking slightly yellow (not shown because it is difficult to photograph the Sun and exhibit its true color, since it is so bright compared to the rest of the scenery).

Coming down to sea level (situation (c) of Figure 2-18), sunlight must penetrate much more air: this scatters more blue light, but now also some green light (shown as blue and green arrows). They combine to give the brighter blue sky color that we are so familiar with in mid-day. The loss of some green light makes the Sun now look a tiny bit orange, in addition to yellow.

In late afternoon (situation (d) of Figure 2-18), the sunlight must travel through even more air, as it slopes down more parallel to the Earth's surface. The air now scatters away most of the blue light, much

Figure 2-19: The sky seen from a jet airplane at high altitude: sunlight coming from above penetrates less air, giving a deep-blue sky color. However, much scattered light comes from near the horizon, which has a longer stretch of atmosphere, thus producing more light intensity.

of the green light (and also some red light, as shown with arrows): that leaves the Sun color to be orange. The broader mix of red, green and blue in the scattered light makes the sky color turn to a more pastel-like whitish blue.

As we approach sunset (situation (e) of Figure 2-18), most of the Sun's blue light and part of its green light have been scattered away during their long path through the air, leaving only pink and orange colors in the sky. This is illustrated in Figure 2-20. Haze and clouds consist of water droplets that strongly scatter light, without much changing the light's color (their droplet size is usually large enough that the blue light does not get scattered more than the green or red light).

After sunset (the Sun sinks below the horizon), the sky color becomes even redder and darker: see situation (f) of Figure 2-18.

How about the colors of clouds, haze, fog, rain, dust and smoke? These can add many colors to the sky, mainly by scattering sunlight (think of the rainbow, Section 2.5). One example after sunset, when red and orange dominate against a weakly blue sky, is shown in Figure 2-21. Another example before sunrise shows more pink colors and "sunrays" that are usually shadows cast by some clouds onto other clouds: Figure 2-22.

Figure 2-20: A low Sun seen through a hazy sky (with reflection from a bay). The Sun is strongly overexposed, incorrectly making it appear white (by also exciting the green and blue sensors in the camera).

A word of caution: don't compare colors between different photographs closely! As mentioned before, cameras "optimize" colors automatically, giving different results in different conditions. In particular, these pictures were taken many years apart with different cameras in different places, some with a visible Sun, some without visible Sun, so the colors cannot be compared directly. Photographs are normally supposed to be more pretty than accurate!

Clouds have many different shapes and densities. The small or thin clouds (as in Figure 2-21 and Figure 2-22) are often largely transparent; they scatter incoming light in all directions, like haze. Larger dense clouds, containing more water droplets, are relatively opaque: light does not penetrate them far, so they reflect sunlight as bright white light in daytime, or become orange or red near sunset and sunrise, reflecting the Sun's and sky's colors; when their water content is high, for example right before they drop rain, clouds look menacingly black away from the Sun, when they are in their own shadow (see Figure 2-23).

Many more spectacular effects can be seen in the sky if one looks carefully, beyond the rainbow: they include dew bows, halos, sun dogs and sun pillars, coronas, cloud iridescence, glories and heiligenschein, mirages and fata morganas, star twinkling and the Sun's green flash at sunset, and more! Some of these are due to ice crystals in clouds, as opposed to water droplets: crystals can act as little mirrors or prisms,

Figure 2-21: A cloudy sky seen from sea level after sunset.

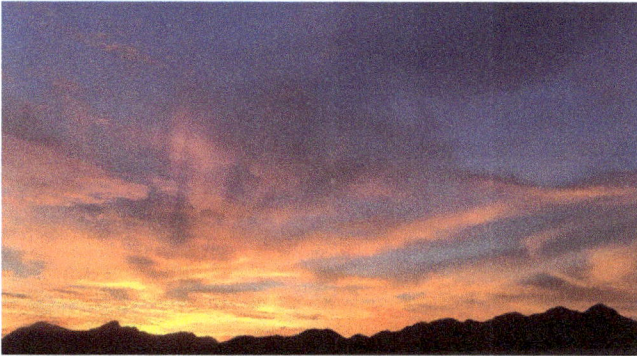

Figure 2-22: Sunrays before sunrise.

Figure 2-23: Opaque dark and white clouds with a rainbow. The Sun is behind the camera.

producing intricate light shows. These effects are described in wonderful books by M. Minnaert[3] and Robert G. Greenler.[4]

If we climb up again by airplane, the sky after sunset can also become gloriously colored, as seen in Figure 2-24. We now see essentially all the colors mentioned above, from almost black to deep red, in one single view.

Figure 2-24: Photograph of the horizon (and the Moon) after sunset from high altitude in a jet airplane: the various colors depend on the path length of sunlight through the atmosphere.

2.9 What have we learned in this Chapter?

We have seen that there are many colors, in fact an infinite number of colors. We also realize that colors are not uniquely named, nor do they look the same when seen by different people or different cameras or in varying lighting or viewing conditions.

We have also learned that we can use just three basic colors (such as red, green and blue, which are called the primary colors) and varying intensities to create an infinite number of colors that our eyes can detect.

---

[3] M. Minnaert, *"The Nature of Light and Colour in the Open Air"*, Dover, New York, 1954, ISBN-10: 0486201961.

[4] Robert G. Greenler, *"Rainbows, Halos, and Glories"*, Cambridge University Press, Cambridge, 1999, ISBN-10: 0521236053.

Black is the absence of light, due to absence of light sources or complete absorption of light. White is primarily the color of sunlight, which is composed of many colors (hues) in the solar spectrum. Rainbows give a slightly smeared out version of the solar spectrum. White can also be seen as an equal mix of red, green and blue. Gray is weak white (or weak black), and is composed of equal amounts of darker red, green and blue.

We define an object's color as the color it shows under illumination with white light.

Complementary colors exhibit strong contrast that can be used to draw attention or excite. They also play a role in explaining the colors of Sun and sky throughout the day.

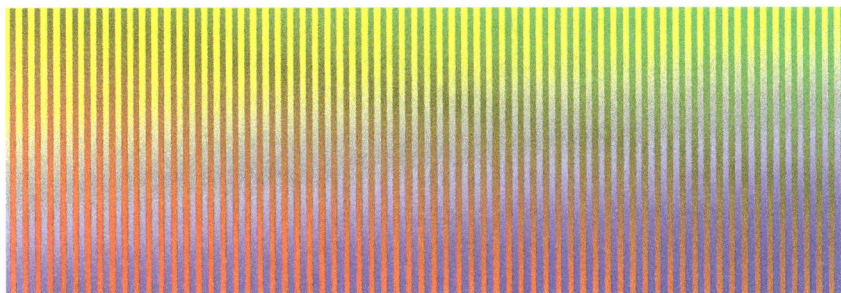

3

The Full Cycle of Imaging with Colors

When looking at an image, we often wonder how "faithful" it is, and in particular whether it has been intentionally changed ("photoshopped"). There may be gross changes, such as modified identities or shapes of people in a photograph, or more subtle changes, such as adjustments of the colors. In photography, as well as in moviemaking, many transformations occur between the original scene being photographed or filmed and the scene perceived in our minds. Each transformation is an opportunity for changing the content and appearance of the images, whether intentionally or involuntarily. These steps include collecting light, recording and storing images, editing and displaying them, and interpreting them in the brain. The resulting mental images can look quite different from the original scene.

— ⚬⟩⟩ ⟨⟨⚬ —

Does a picture faithfully display the original scene it represents? In particular, are the colors of the image correctly reproduced? When we look at a color picture or a color movie, we rarely think of all the transformations that happened between the original scene and the finished picture or movie. As we will discuss in more detail in later sections, modern digital recording of a picture converts the scene using a number of automatic or selected transformations by the camera lens and detector system: the resulting recorded images are rarely identical in color to the original scenes, so here we already lose some degree of "fidelity", mostly involuntarily. Pictures meant for publication are later often intentionally "enhanced" or "improved" ("photoshopped", for example to remove the "red-eye" effect). Next, the display of images for viewing depends on ink fidelity for accurate printing or on reliable emitter materials for displaying on computer screens: here again, colors can easily change, most often involuntarily. The same is true for **movies**, where in addition sound plays a role and a storyline is important.

This **cycle of imaging** is schematically depicted in Figure 3-1, which starts with the original scene at top left: this scene includes objects and action, lighting effects, as well as, in movies, desired sounds and undesired noise. The full processing cycle will end up at top right with the image or movie that our brain has prepared for our consumption.

Let's go through this cycle step by step. In later sections, we will describe in more detail many interesting aspects of these various steps.

The first step is to collect the light and sound so as to deliver them to the **detectors** of light and sound. This requires a **lens** system (see Section 5.1.3), which must focus on the desired depth of field and can zoom in or out as well as filter out undesired **ultraviolet light** or minimize light **polarization**, for example; **stereoscopic** viewing may be planned to give a stronger 3-D impression, requiring two slightly different views of the scene. For **movies**, suitable microphones have to be selected and properly placed to optimize sound quality: one or more microphones may be used.

Recording images comes next, described in more detail in Chapter 5. The light and sound can now be recorded by the **detectors**. With **film photography** (still often used for **movies**), the correct exposure times and developing chemicals are needed to obtain good colors. With **electronic recording**, one must pay attention to **pixilation** (which breaks up a picture into tiny picture elements called **pixels**, and which can create artificial **moiré patterns**, for example, see Chapter 9). Also needed is faithful **recording of color**, while adjusting the white balance

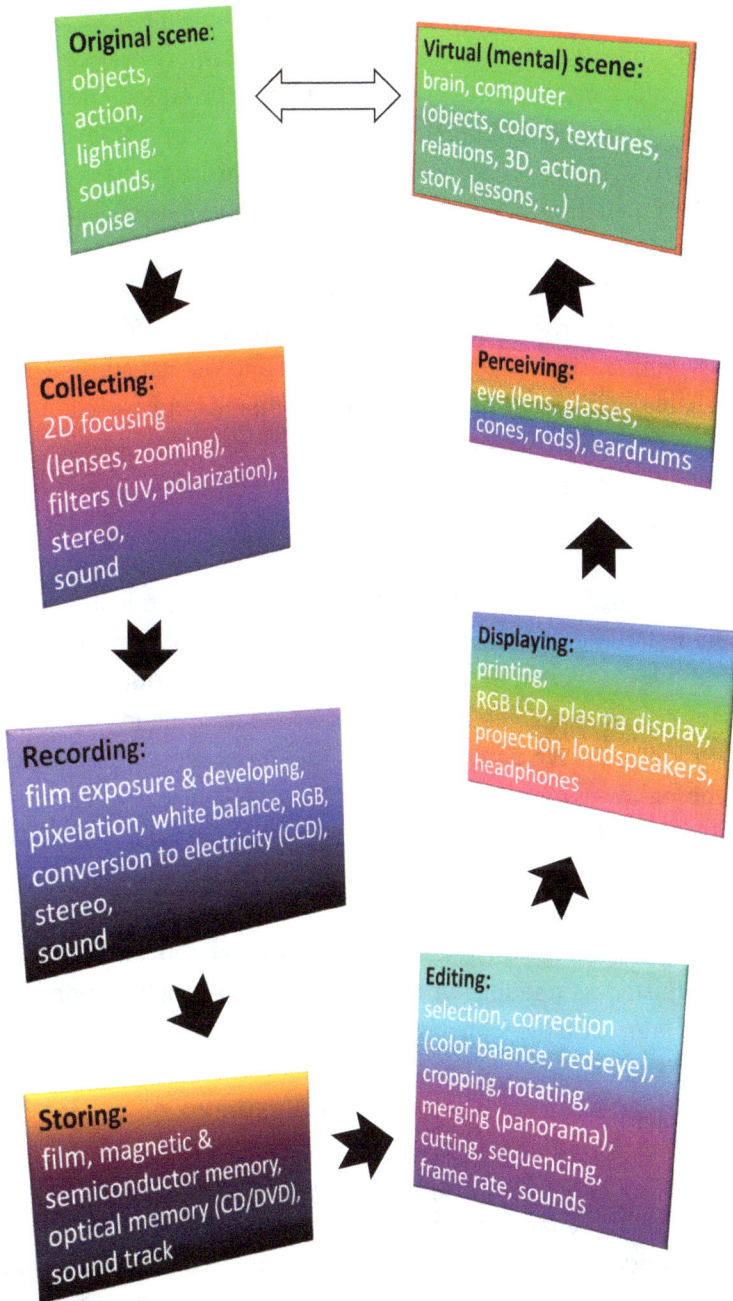

Original scene: objects, action, lighting, sounds, noise

Virtual (mental) scene: brain, computer (objects, colors, textures, relations, 3D, action, story, lessons, ...)

Collecting: 2D focusing (lenses, zooming), filters (UV, polarization), stereo, sound

Recording: film exposure & developing, pixelation, white balance, RGB, conversion to electricity (CCD), stereo, sound

Storing: film, magnetic & semiconductor memory, optical memory (CD/DVD), sound track

Editing: selection, correction (color balance, red-eye), cropping, rotating, merging (panorama), cutting, sequencing, frame rate, sounds

Displaying: printing, RGB LCD, plasma display, projection, loudspeakers, headphones

Perceiving: eye (lens, glasses, cones, rods), eardrums

Figure 3-1: Processing cycle of an image or movie from original to perceived scene. (The colors here have no physical meaning.)

to make sure that, for instance, white does not appear as pinkish or bluish off-white (see Section 6.2.1). A digital camera will convert the received colors into electric currents for each pixel and then into so-called RGB components in digital form, such as intensity levels from 0 to 255, as described in Chapter 4. One or more microphones will convert sounds into digital form, either mixed together or kept separate for later manual mixing.

The third step is **storing the image**: storage of the recorded information for later processing. Film must be protected from scratching, humidity, *etc.* Various electronic, magnetic and optical storage devices are available, with good error correction schemes in case bits of information accidentally change over time.

Next comes **editing**. We can change many aspects of the stored image or movie. First, we select the "best" images among the many we may have taken. Even so, they may be imperfect and need corrections or modifications: we may want to **balance** the colors toward "warmer" (redder) or "cooler" (bluer) colors (Section 5.3), and we may remove any "**red-eyes**" (Section 5.1.3), ugly telephone lines or untimely objects (like modern wrist watches and smartphones in a scene that we want to look medieval). We may adjust the framing of the scene by zooming in, **cropping** away undesired edges or tilting the picture. We may also assemble a wide panorama by merging multiple images. In a movie we may cut and reorder sequences of frames. The frame rate may need to be adjusted for different presentation formats (such as for movie theatres *versus* television *versus* online viewing). Similarly the sounds can be adjusted as needed. Of course, the result should now be stored again for later display. Copying may be needed for sharing, which is delicate with film.

Displaying an image or movie includes **printing** on paper or other materials, **displaying** on a computer monitor or television screen or large advertisement board, and projecting onto a screen (Chapter 6). Any of these devices can change the apparent color to some extent: colors are rarely reproduced perfectly faithfully in print or on an electronic display, due to the materials used. Sounds may depend on whether they are meant for the open air or a closed room or headphones; they may also need to be stereo or monaural sounds.

Next our eyes and ears perceive the images and sounds. The eye's **lens** focuses the incoming image on the **retina**, with more or less success, perhaps with the help of **corrective glasses**: sharpness may be imperfect. In the retina, our **cones** and **rods** (Section 5.1.1) convert the light into signals that are sent over nerves toward the **brain**: each observer will see the image slightly differently (Section 5.1.5). Our eardrums similarly convert incoming sound into signals for the brain, also with less than perfect fidelity.

In the last step, our **neural system**, which mainly includes our brain, first interprets the signals sent by the eyes as "red", "green", "blue", "gray" or combinations thereof. It then analyzes the information received to identify objects, with their colors and **textures** (such as surface roughness), as well as their three-dimensional positions and relationships (and their motions in movies). The brain reconstructs the original scene in the form of a virtual or mental version of that scene, as we discuss in Chapter 10. It also monitors the actions and the stories behind what it sees and hears. The brain usually also remembers stories and learns lessons from what it saw and heard.

The bottom line is that the perceived image is different from the original scene in many ways. Thus, no image can be assumed to be exactly faithful to the original scene. The largest changes occur in the recording and displaying steps, mostly involuntary, despite the best efforts of the designers and engineers of recording equipment. Significant changes can also occur in the **editing** step, mostly intentionally, and in the brain, mostly involuntarily.

Fortunately, we are rather tolerant of changes in colors. There is a good reason for that: the lighting in our daily lives often changes dramatically, frequently modifying the apparent colors of familiar objects. Think of the sunlight changing from yellow/orange at sunrise to white at noon and orange/red at sunset, while clouds can change the illumination to light or dark gray. Also, when we go indoors, the lighting can be reddish (warm and gentle) in a comfortable living room, but white (harsh and violent) in a bathroom or office, for example. Simple shadows can also strongly change the appearance of objects. Life would be very difficult if we had to reinterpret what we see whenever such lighting changes occur. Thanks to our remarkable brain, we are used to such changes and hardly notice them (see Section 11.1).

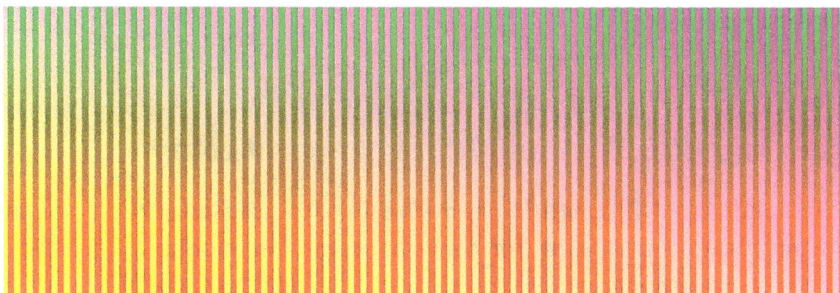

4

Describing and
Working with Colors

*In Chapter 2, we have discussed what **colors** we see and how they are related. We wish to next discuss more practically how we can systematically describe colors. This will then allow us to address how we can record colors with our eyes and with photography, and how we can make colors with ink, paint and electronic displays such as television sets and computer monitors.*

With sunlight we can create all human-observable colors by suitable mixing of its infinitely many spectral colors pictured in Figure 2-12 or simulated in Figure 2-11. Fortunately, as we have seen, there is a much simpler way to describe almost any color we wish, by mixing just three primary colors instead of an infinite number of colors. The reason is that our color vision depends on only three types of cones in our eyes.

We can therefore simplify the description — and also the production — of almost all visible colors by using just three basic colors. We will start by explaining three systems of colors which achieve that goal and are much used in practice: the so-called RGB, CMY and HSL systems of color. These systems will allow you to easily produce millions of colors on your computer.

⇐·⟩⟩ ⟨⟨·⇐

4.1   The RGB, CMY and HSL systems of colors

The colors red (R), green (G) and blue (B) are used as primary colors in the electronic industry for cameras, computer monitors, television screens, and other color displays. Figures 2-5 and 2-13 show these colors and some of their mixtures. As we will see, the intensity (brightness) of the resulting mixed color can also be varied. This approach is called the **RGB color system**.

The colors cyan (C), magenta (M) and yellow (Y) are preferred in the printing industry for clothing, inkjet printers and other applications: together with their mixtures and intensities, they form the so-called **CMY color system**, illustrated in Figure 4-1.

What is the relation between the RGB and CMY systems of color? Compare the two plots with overlapping circles for the CMY and RGB color systems, shown in Figure 4-1. With CMY, the large circles are

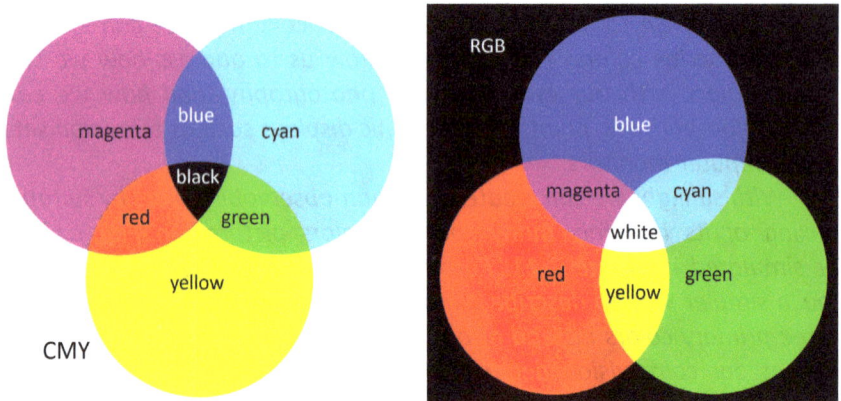

Figure 4-1: Left: Overlapping cyan, magenta and yellow circles, mixing to form red, green and blue petals, and a black center. This construction uses the subtractive CMY system of colors, starting from a white background. You can imagine looking at a white sheet of paper through three circular filters colored cyan, magenta and yellow: where you look through only one filter you see cyan, magenta or yellow; where you look through two of these filters you see red, green or blue; and where you look through all three filters you see nothing, namely black. Right: RGB color system, as in Figure 2-13.

colored cyan, magenta and yellow: we already saw these three colors in the RGB system, where red, green and blue overlap. Conversely, in the CMY system, cyan, magenta and yellow combine pairwise to produce red, green and blue that are the primary colors of the RGB system (they are the colors of the circles in the RGB plot): see Figure 4-1.

We face two puzzles here. In Figure 4-1, the colors red, green and blue are clearly less bright than the colors cyan, magenta and yellow: this does not look like addition of light in the CMY system; adding light should brighten, not darken! Also, the overlap of the three CMY colors in the center gives black, not white: why? Let's address these questions next.

How can combining colors create black? A clue lies in the surrounding background color, which is black for RGB, but white for CMY. Remember that RGB is used for computer displays, *etc.*, while CMY is used for printing, *etc.* A computer display is black when turned off, because it emits no light (actually, such a display also reflects some light and is thus not completely black, but let's imagine using it in a dark room): that is the black background color shown in Figure 2-13. Colored light has to be <u>added</u> to that black background for a computer display to emit light. We will describe how this can be done in practice in Section 6.2.

Thus, RGB is an <u>additive</u> process, in which light of different colors is added to a black background.

The reverse is true of printing and painting. There we normally start with a white background: the paper or canvas only <u>reflects</u> light, without emitting light. We must assume that white light falls on the paper or canvas, for example sunlight. Since the reflected white already includes a maximum amount of color, we need to <u>subtract</u> color from the white light reflected by the paper or canvas. The ink or paint does exactly that: it acts like a filter that subtracts color during the reflection of white light. The ink or paint filters out unwanted colors from the white light, leaving only the desired color. How to do this in practice will be discussed in <u>Section 6.1.</u>

Thus, CMY is a <u>subtractive</u> process, which removes color from a bright white background, as shown in Figure 4-1.

What is the HSL color system? The HSL system is an alternative approach to describing colors. It is more intuitive than RGB and CMY: it is indeed a convenient way to think of and create colors, as we will see in more detail in Section 6.2.3. This system organizes colors by **hue** (H), **saturation** (S) and **luminance** (L), hence the name **HSL**.

When you think of a color, you probably first think of the hue, described in Section 2.2. The **hue** is the color that we see along the outer edge of the color triangle (see Figure 2-5): it ranges continuously from red through magenta, blue, cyan, green and yellow back to red. The **luminance** is simply the intensity or brightness of a color; for example, the hue cyan could be more or less luminous.

The **saturation** of a color describes how much gray is mixed into the color, as if looking at it through gray fog. This needs more explanation. Looking at the color triangle of Figure 2-5, we notice that the colors become less "colorful" and more gray as we move towards the center of the triangle (it is like looking through increasingly dense fog near the center of the triangle); at the center itself, the color is in fact "colorless" gray: there the hue is no longer visible. Saturation describes how colorful and gray-free is the color. Thus: A color is saturated when it contains no gray and lies on the outer edge of the color triangle; a color has low saturation when it contains much gray; and gray has zero saturation.

IF YOU HAVE INTEREST IN A GEOMETRICAL REPRESENTATION OF COLORS, ESPECIALLY WITH A 3-DIMENSIONAL GEOMETRY, YOU MAY LEARN HERE ABOUT AN ELEGANT WAY TO ORGANIZE AND UNDERSTAND COLORS IN A 3D MODEL — THE COLOR CUBE:

We have seen that we can produce most visible colors with primary red, green and blue colors, namely the RGB colors. Each of these primary colors can have different intensities or brightnesses from zero up to a maximum value (we set maximum values to avoid blinding our eyes). You may recognize the similarity to a three-dimensional geometrical coordinate system xyz: we can choose x to tell the brightness of red, y the brightness of green and z the brightness of blue. In general in geometry, xyz can represent any three variable quantities, such as: length, width and height, or latitude, longitude and altitude, to specify a position in three-dimensional space; or wealth, age and health to detail quality of life; or US dollars, euros and Chinese yuan to quantify cash wealth.

Look at the cube at left in Figure 4-2. In the bottom left corner we start with no light, namely black: it has zero red, so x = 0, zero green, so y = 0, and zero blue, so z = 0; we therefore label that point xyz = 000. To add red light we follow the cube edge along line x that turns from black to red and ends in corner R00 where we have a maximum amount of red; the label R00 indicates that no green or blue is mixed in with the red: this point represents primary red, while the line from 000 to R00 gives all shades of red, ranging from black with zero brightness to primary red with full brightness.

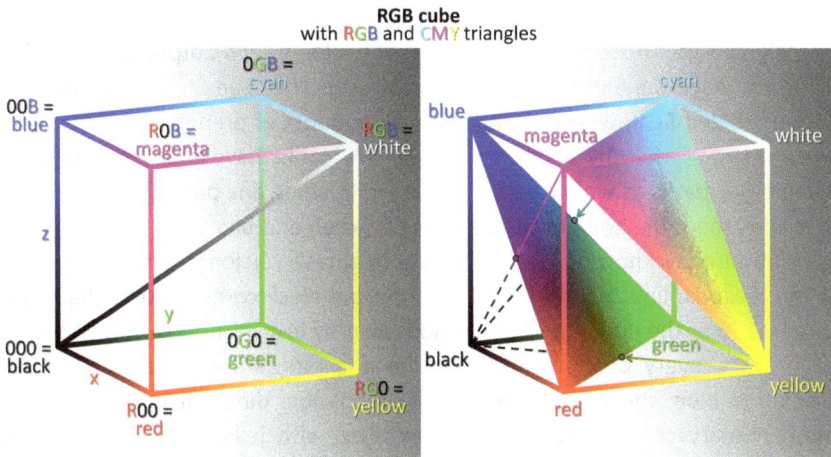

RGB cube
with RGB and CMY triangles

Figure 4-2: Many visible RGB colors are shown here in three dimensions. Pure red follows the x-axis, pure green the y-axis, and pure blue the z-axis, from zero brightness at the point 000 (black) to maximum brightness at points R00 (primary red), 0G0 (primary green) and 00B (primary blue), respectively. These three points form three corners of the cube shown. The corners 0GB, R0B and RG0 combine two primary colors: they represent cyan (which combines green and blue), magenta (red and blue) and yellow (red and green), respectively. The last corner RGB combines all three primary colors and represents white. The gray diagonal in the left cube shows all gray levels from black to white. All points within the cube are combinations of red, green and blue. The two color triangles in the right cube are slices that show some of these combinations: they are the same two triangles shown in Figures 2-5 and 2-8.

Likewise, we could start from the black corner 000 and add green by going along the cube edge y to the green corner 0G0: this line gives all shades of green from black to primary green, without red or blue. Or we could go along the vertical cube edge z to the blue corner 00B: this line gives all shades of blue from black to primary blue, without red or green.

Next, we can start from red (R00) and add green to it: we then follow the cube edge parallel to line y that ends up in the point RG0 at bottom right (which represents yellow); you can see along that edge how red turns gradually

(*Continued*)

(Continued)

from primary red *via* orange and brownish yellow to primary yellow. Remember that basic yellow can be made as an equal mix of primary red and primary green, hence the label RG0! We could have reached the same point RG0 (yellow) by starting from the primary green point 0G0 and adding red, parallel to the line x. Similarly, we could travel along other edges of the cube, to reach corners R0B (magenta) and 0GB (cyan).

We can go further: starting from RG0 (yellow), let's add blue by moving along the edge parallel to z (vertically upward). This leads to the upper right corner labeled RGB, where we have equal amounts of primary red, green and blue: we have thereby generated white. We could have reached this point along other paths as well: we also obtain white along any path within the cube that adds the same amount of primary red, green and blue.

You already know that white is the brightest version of gray. Now look at the diagonal line that goes straight from the black corner 000 to the white corner RGB of the cube: this line includes all gray levels from black to white.

In fact, every point within the cube is a different combination of red, green and blue; these points represent all possible different mixtures of the three primary colors. Imagine the cube filled with jelly, such that the jelly has a different color at every point in the cube. It is not easy to graphically show all those points within the cube, but we can show some flat slices cut through the inside of the cube, as if cutting slices of jelly. The right image in Figure 4-2 shows two such triangular slices. The left slice connects the corners representing the red, green and blue primary colors: this slice is none other than the color triangle which we showed in Figure 2-5! Notice how the colors of this slice match up exactly with those of our earlier color triangle.

Furthermore, consider the slice drawn through the corners for cyan, magenta and yellow (C = 0GB, M = R0B and Y = RG0). This CMY slice has the exact colors which we saw earlier when we brightened up the center of the RGB triangle, see Figure 2-8, as suggested by the arrows in the right image in Figure 4-2. In fact, the colors in the CMY slice are double the brightness of the corresponding colors in the RGB slice, because they are twice as far from the black corner (their xyz values are therefore doubled).

In the perspective views of Figure 4-3, we see only the faces of the cube. The left view looks down into the black corner of the cube (whose front and interior are here assumed transparent and the cube is shown hollow): so the black corner is in the center of this view and we see only the three darker back faces of the cube. The right view looks in the same direction at the outside of the cube (now not transparent): it looks down at the white corner, so the black corner is out of sight and we see the three brighter front faces of the cube.

RGB cube: 6 faces

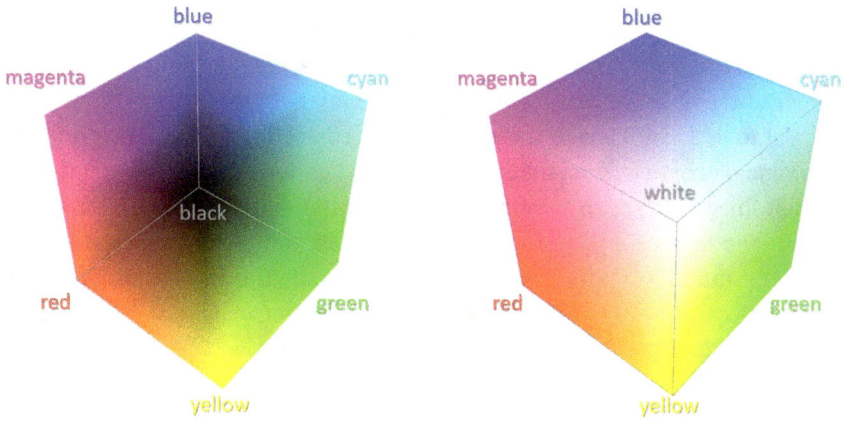

Left: looking down into the black corner   *Right*: looking down at th white corner

Figure 4-3: Views of the six faces of the color cube of Figure 4-2. The three darker faces of the cube are seen at left (using a hollowed-out cube), the three lighter ones at right (with a filled cube). The six edges have exactly the same colors in the left and right views. (*Source*: Courtesy of K.E. Hermann.)

RGB cube (looking down at white corner): **square slices**

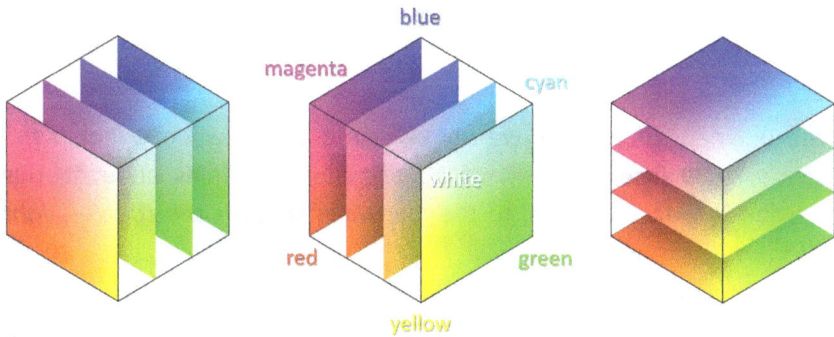

Left: each slice has constant red (back slice has 0% red; front slice has 100% red)
Center: each slice has constant green (back slice has 0% green; front slice has 100% green)
Right: each slice has constant blue (back slice has 0% blue; front slice has 100% blue)

Figure 4-4: End-on views of the color cube of Figures 4-2 and 4-3. Each slice has a constant brightness of one primary color: going from front to back, the middle slices have 2/3 and 1/3 of the maximum amount of red, green or blue, respectively. (*Source*: Courtesy of K.E. Hermann.)

(Continued)

(Continued)

We can make another attempt at looking at the variety of colors inside the cube: see the three images in Figure 4-4. We are again looking at the white corner of the cube (at front center of each cube), but we now only draw slices parallel to its faces. Each of these slices has a constant amount of one primary color and all possible amounts of the other two primary colors. For example, in the left image we see in front a slice that has the maximum possible amount of red, and in back a slice that has no red at all: the two intermediate slices have 2/3 and 1/3 of the maximum amount of red. The middle image shows slices with constant amounts of green (maximum green in front and no green in back), while the right image has slices with constant amounts of blue (maximum blue in front and no blue in back).

Warning: there is nothing "cubic" in our eyes! The cube used in describing RGB colors is a purely theoretical model: it's a mathematical model. The cube is very convenient to represent and understand the relationships between colors, but has no reality in itself. This is a nice example of models and theories in physics and science more generally: the model helps us to understand, calculate and even predict reality, but is only a mental substitute for that reality. It is one of the mysteries of nature why it is even possible to dream up such models that "mimic" nature so well.

4.2 Reflected colors

With our new knowledge, we can return to the question: **What is the color of a surface that reflects light?** When you think about it, this question is in fact identical to the following: **What is the color of light reflected by a surface?**

The equivalence of these two questions is not obvious: as we discussed in Section 2.6, we have the intuitive feeling that any object has a fixed color of its own, independent of the lighting. For example, we may think that a red ball will always be red. However, a purely red ball under purely blue light will actually be black, because that ball will not reflect the blue light and there is no red light that can be reflected by that ball. More surprisingly: a cyan ball under yellow light will look green, because yellow light contains green light that can be reflected by the ball, while the red light in yellow cannot: this result will become clear in this Section.

I already mentioned in Section 2.6: **When we speak of an object's color, we actually mean its color in white illumination.** This is because we normally view objects in white light, so we assume that the color we see belongs to those objects and does not depend on the incoming light. Figure 4-5 represents that case: white light falls on surfaces of different colors; the reflected light is then simply the surface's own color. This can be understood by decomposing the white light into its red, green and blue components and by tracking which of those components is reflected. A color component that is not reflected is removed by being **absorbed**: it disappears into the surface, usually in the form of heat, warming up the surface.

Absorption of color is governed by the chemical composition of a surface. Most chemicals absorb a part of the visible light, such as red, green, blue or any combination of them; those chemicals that don't absorb any colors are transparent, such as glass and water. Exactly which color or colors are absorbed by a given chemical is determined by quantum mechanics, a subject in physics and chemistry that is beyond the scope of this book.

As illustrated in Figure 4-5, a white surface reflects all colors, so red, green and blue are all reflected, and the reflected light is white. A black surface reflects no colors, so there is no reflected light. A red surface reflects only red light, while a cyan surface reflects both green and blue, giving a cyan reflection. This is easily understood by tracking separately the red, green and blue light components.

Now try to imagine replacing the white incoming light falling on a colored surface by colored incoming light: what do you think will happen? Again it helps to decompose the incoming light into its red, green and blue components and to follow each color component separately as it reflects from the colored surface. This is illustrated in Figure 4-6. After the reflection, all that remains to do is to reassemble the reflected color components to predict the resulting combined reflected color.

Now, red incoming light will reflect off any surface that contains a red component, whether it is white or red or magenta or yellow. But, if the surface has no red component, so it could be green, blue, cyan or black, then red cannot be reflected and the surface looks black under red illumination. With yellow light (containing red and green) falling on a cyan surface, only the green component of the yellow light can reflect

Reflected colors: <u>white</u> incoming light

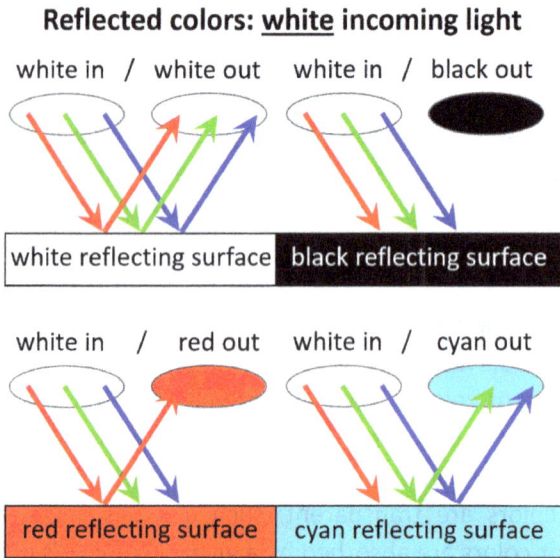

Figure 4-5: Reflection of white light by surfaces of various colors. The red, green and blue components are tracked separately.

Reflected colors: <u>non-white</u> incoming light

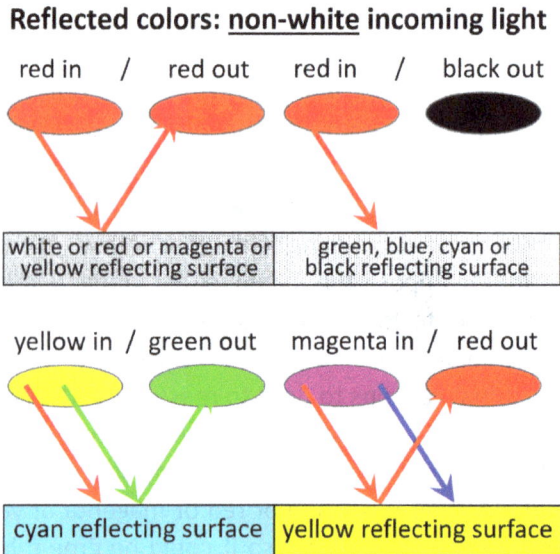

Figure 4-6: As Figure 4-5, but for reflection of non-white light by surfaces of various colors. The red, green and blue components are tracked separately.

from the cyan surface, because cyan has a green component but no red component. And magenta light only has red in common with a yellow surface, so that only red light will be reflected.

The cases illustrated in Figures 4-5 and 4-6 are brought together with other cases in Figure 4-7: it shows a **table of the reflected colors** which you can expect to see when shining colored light onto a colored surface. This table is simplified: it assumes that we have only the "pure" basic RGB and CMY colors for both the surface and the incoming light; we also ignore further absorption of light by the surface, thus resulting in "pure" reflected colors.

The table of Figure 4-7 also includes the effect of gray colors. With gray incoming light or a gray surface color (here labeled with lower case as "r + g + b", meaning "weak white"), only part of the intensity of white comes in or is reflected, so darker versions of the basic colors are reflected; when gray light falls on a gray surface, only a part of the gray light is reflected, giving an even darker gray, labeled in the table with italics as "*r + g + b*" instead of "r + g + b".

How was the table of reflected colors, *Figure 4-7*, obtained? As suggested above, I first decomposed both the incoming light and the surface color into R, G and B components; then, for each R, G or B component of the incoming light, I checked whether it is reflected by the surface. For example, if both the incoming light and the surface contain red, then red light will be reflected, otherwise no red light will be reflected. And I checked likewise whether green and blue are present and reflected: only if the surface color has a green component will it reflect green light, and similarly for blue. Thus: **A reflected color consists of the RGB components that are common to both the incoming light and the surface.**

Note the following symmetry, visible in Figure 4-7: you get the same result if you interchange the incoming color and the surface color. For example, yellow light falling on a magenta surface gives the same reflected color (red) as magenta light falling on a yellow surface.

In reality, the incoming color and surface color will rarely be as simple as shown in Figure 4-7. Not only can both colors be any combinations of red, green and blue, but there will also normally be a color dependent loss of intensity because reflection can be partial, not only 0% or 100%. Figure 4-7 will nonetheless give an approximate idea of the reflected color.

Incoming light color								
Surface color	White (R+G+B)	**Red** (R)	**Green** (G)	**Blue** (B)	**Yellow** (R+G)	**Mag-enta** (R+B)	**Cyan** (G+B)	**Gray** (r+g+b)
White (R+G+B)	white (R+G+B)	red (R)	green (G)	blue (B)	yellow (R+G)	mag-enta (R+B)	cyan (G+B)	gray (r+g+b)
Red (R)	red (R)	red (R)	black	black	red (R)	red (R)	black	dark red (r)
Green (G)	green (G)	black	green (G)	black	green (G)	black	green (G)	dark green (g)
Blue (B)	blue (B)	black	black	blue (B)	black	blue (B)	blue (B)	dark blue (b)
Yellow (R+G)	yellow (R+G)	red (R)	green (G)	black	yellow (R+G)	red (R)	green (G)	dark yellow (r+g)
Mag-enta (R+B)	mag-enta (R+B)	red (R)	black	blue (B)	red (R)	mag-enta (R+B)	blue (B)	dark mag-enta (r+b)
Cyan (G+B)	cyan (G+B)	black	green (G)	blue (B)	green (G)	blue (B)	cyan (G+B)	dark cyan (g+b)
Gray (r+g+b)	gray (r+g+b)	dark red (r)	dark green (g)	dark blue (b)	dark yellow (r+g)	dark mag-enta (r+b)	dark cyan (g+b)	darker gray (r+g+b)

Reflected color

Figure 4-7: This table shows reflected light colors resulting from incoming light (with color shown along the upper edge) after reflection from a surface (with color shown along the left edge). For example, yellow incoming light (4th vertical column from the right) reflects from a cyan surface (one but last row near the bottom) to give green light (at the intersection of that column and that row). Here, "dark" colors are defined as having half the intensity of the normal color, so that "dark gray" has half the intensity of white, and "darker gray" has a quarter of the intensity of white.

Do metallic and other shiny surfaces have more complex colors than we have described so far? The light reflected by many automobiles, guitars and glass objects, among others, is shiny. In some cases, the reflected color of paint actually changes as you look at it from different angles. There are also sparkling paints, in which tiny points of light brighten and dim as you move, similar to the glitter of crystal reflections. Four main effects are at work here.

The **shine** or **gloss** is due to a very smooth surface: if you scratch a smooth metal surface it will soon lose its shine and look as dull as a gray piece of paper. Rust will do the same as scratches, which is why gold keeps its shine, since it does not rust (also, since gold is soft it is easily smoothed to a nice shine, even after being scratched). Non-shiny surfaces reflect light in all directions due to their rough surface; good examples are most types of cloth and non-glossy paper, as is the **matte** paint of some luxury cars and old unwaxed cars. Plastics are shiny or matte depending on their texture: smooth or rough, respectively. Physicists use the term **specular reflection** for **shine** (like a mirror), contrasting with **diffuse reflection** into all directions (like matte surfaces such as non-glossy paper).

The **sparkle** of some paints is due to very small metallic flakes embedded inside a slightly transparent paint. The flakes have flat surfaces which act as small mirrors. If the flakes are lined up well, the little mirrors will reflect light in some preferred directions and not in others, creating the sparkle as you move around.

The color variation with viewing direction that occurs in some paints is due to the wave character of light. This effect, which is based on both **refraction** and **interference** of light, also uses tiny flakes inside paint, but in addition uses the light-splitting effect seen in the spectrum of Figure 2-9 created by a prism or in rainbows (see Section 2.5), as well as interfering reflections from different surfaces. As with a prism or a rainbow or a thin film of oil on water, the color that you see changes as you move around.

These effects — shine, sparkle and changing color — can be reproduced by RGB and CMY colors, but only in a fixed view. RGB and CMY colors will not change with viewing direction, so that a shiny spot or sparkle will not change as you move around an image; however, a movie will show the changing shine or sparkle as the camera moves around the original object (car, guitar, *etc*.), because each frame is a new image taken from a different angle, just as if you moved your eye around the object.

WE HAVE LITERALLY BEEN ADDING AND SUBTRACTING COLORS. CAN WE REALLY DO THAT AS SIMPLE SUMS AND DIFFERENCES (SIMILAR TO CURRENCIES)?

In Figure 4-7, we added "R + G" to make yellow and "R + G + B" to make white. In a similar way we may also remove colors, as we do in discussing reflection from a colored absorbing surface. For example, if we take white light (R + G + B) reflected from a yellow surface (R + G), which removes blue (B), we are left with R + G + B − B = R + G, which is yellow as expected. Or if we take magenta light (R + B) reflected from a cyan surface (G + B) that removes red (R), we get R + B − R = B, which is blue as expected.

Thus, we can add and subtract colors as if we were dealing with money of different currencies: we just have to make sure that we treat the different primary colors red, green and blue separately, just as we would treat different currencies separately (we can't "subtract" Russian rubles from British pounds, and we can't "add" 100 dollars + 100 euros). We can make this approach a little bit more systematic with the notation R + B + G = "RBG" for white, R + B = "R0B" for magenta, G + B = "0GB" for cyan, R = "R00" for red, B = "00B" for blue, *etc*., which we already used in Figure 4-2: now each color is represented by a set of three values (similar to dealing with three currencies); the first value is red, the second green and the third blue. Here "0" simply means a zero amount of the corresponding primary color, so that "000" means no color at all, namely black.

With this notation, the above example R + G + B − B = R + G becomes: RGB − 00B = RG0. The rule is quite simple: treat each color separately, namely: R − 0 = R (no red is removed), G − 0 = G (no green is removed), B − B = 0 (blue is removed), giving altogether RG0. Similarly, the example R + B − R = B then becomes: R0B − R00 = 00B (only red is removed).

We need to add one natural constraint: we don't allow negative amounts of color in the result, since that does not exist; we can say that it is not realistic or "physical" (with money we tolerate negative amounts, namely debts, but nature doesn't tolerate negative colors). Thus, a green surface 0G0 removes red R00 and blue 00B, so that red incoming light R00 is reflected as R00 − R00 − 00B = 000 or black, and *not* as 00−B with a negative amount of blue (since 0 − 0 − B = −B becomes negative, we set it to 0). In words: we cannot subtract a color that was absent in the first place.

We may use smaller amounts of light than the maximum possible, as we do with gray. For example, a magenta surface R0B will remove green 0G0 light from incoming gray light rgb (the lower case here represents half the intensity of white), resulting in the reflected light rgb – 0G0 = r0b or half-intense magenta (strictly g – G would give –g since G = 2g, but we set negative results to zero). A more general notation for this would be: rgb – 0G0 = (0.5, 0.5, 0.5) – (0, 1, 0) = (0.5, 0, 0.5) = r0b (note that 0.5 – 1 is set to zero). Here the numbers 0.5 and 1 could in general be anything in the range from 0 (meaning zero intensity) to 1 (meaning maximum intensity).

Sums and differences of colors also allow us to better understand and describe **complementary colors**, discussed in Section 2.7. Consider the complementary pair red and cyan. From Figure 2-5 we know that: red + cyan = white; we also know the reason, namely that cyan is green + blue, so that red + cyan = red + green + blue = white. We can also change that first relation to cyan = white – red, or red = white – cyan. That is the meaning of complementary: if we remove red from white, we are left with cyan; and if we remove cyan from white, we are left with red; or we can say that red and cyan add up to white and are thus complementary. This can all be written very succinctly as: R00 + 0GB = RGB (for red + cyan = white); 0GB = RGB – R00 (for cyan = white – red); and R00 = RGB – 0GB (for red = white – cyan). These relations can easily be generalized to the other colors of the complementary color circle that lie between the six basic colors.

Some readers may recognize here the so-called **vector notation** in 3 dimensions: a vector in three-dimensional space specifies a position or a displacement in that space given by the three quantities x, y and z. Vectors are often written as (x, y, z) and can be manipulated by addition and subtraction as we did above. For example, moving from the starting position (xstart, ystart, zstart) through a displacement vector (xmove, ymove, zmove) results in a new position (xstart, ystart, zstart) + (xmove, ymove, zmove) = (xstart + xmove, ystart + ymove, zstart + zmove) = (xend, yend, zend). Thus, if you start at position (0, 0, 0) and move east by 3 meters (xmove = 3), north by 2 meters (ymove = 2) and up by 1 meter (zmove = 1), you will end up at the end position (0 + 3, 0 + 2, 0 + 1) = (3, 2, 1) meters.

Readers may also be reminded of the color cube shown in Figure 4-2. Indeed, the three-dimensional cube has three axes x, y and z that correspond exactly to the three primary colors: the vector (x, y, z) is an arrow pointing from the "black" cube corner 000 to any mixed color in the color cube.

4.3 How many colors do we need?

In Section 2.1, you have listed a number of colors: perhaps a few dozen, but probably not over one hundred. This suggests that a hundred or so colors may suffice for all your daily needs: *is that the correct number of colors?* That count, however, may neglect one aspect: the **intensity** of the color. Indeed, the apparent color of most objects varies smoothly across their surface, depending on illumination and point of view, even if that object has a single surface color or hue (we also discussed this in Section 2.1). Look at a ball or a human face: each may have a single "color", but most likely a vast range of reflected light intensities.

This is clearly seen with the photograph of a uniformly white bowl shown in Figure 4-8, sitting in yellowish early-morning sunlight. The 8 versions of the photograph of that bowl have different numbers of intensity levels: from 256 levels (at top left, as photographed) *via* 128 (next down), 64 (down), 32 (bottom left), 16 (bottom right), 8 (next up), 4 (up) to 2 levels (at top right). With 128 levels of intensity, we already can see discontinuous jumps or steps in the color; these jumps become much stronger with fewer levels. For comparison, I also repeat the smooth black-to-white color bar in Figure 4-9.

Another example is the delicate tones of the clear sky: Figure 4-10 shows four versions of one of my pictures taken at sunrise (with a mobile phone camera giving 256 intensity levels). The first version has 256 intensity levels of each color, the second 128 levels, the fourth 64 levels and the last 16 levels: already with 128 levels and even more with 64 levels, we see artificial steps and bands in the sky's color, spoiling the whole picture.

We could try to create a continuous intensity variation, with no jumps or steps at all. However, this could be more expensive in terms of storage size of the picture and would require more complex display equipment. Therefore, we wish to minimize the number of color intensities: a compromise is needed between expense and perceived continuity of color variations.

From the above examples, clearly, **128 intensity levels are not sufficient, while 256 levels seem adequate**.

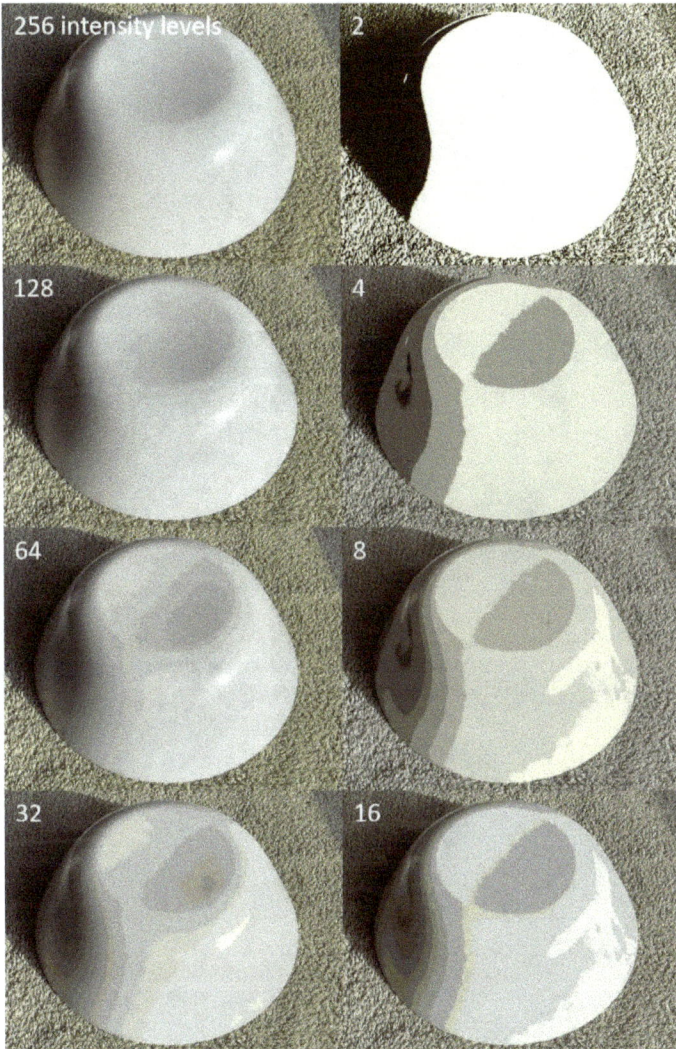

Figure 4-8: A uniformly white bowl in yellowish sunlight shown with 8 different numbers of intensity levels, from 256 levels to 2 levels as marked. (The bowl rests on a greenish bath towel.)

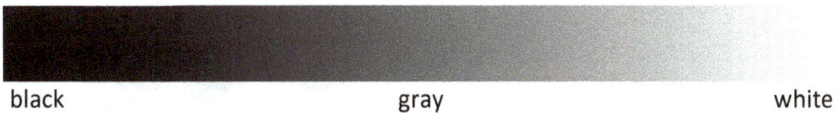

black   gray   white

Figure 4-9: A smooth progression from black to paper white.

Figure 4-10: The same picture of a sunrise is shown with 256 intensity levels (top), 128 levels (second), 64 levels (third) and 16 levels (bottom).

Therefore, how many colors do we actually need? We will need
to use three primary colors (let's assume red, green and blue) to match
our eyes' three types of cones; notice that we thereby avoid using the
multitude of different hues between those three primary colors, since
the hues are simple combinations of the three primary colors.

We also want 256 possible levels of intensity to give smooth color
intensity variations. We will need to combine any one of the 256 red
intensities with any one of the 256 green intensities and any one of the
256 blue intensities.

Do you have any guess how many combinations we can generate
with all these intensities of red, green and blue? The answer is: we get
$256 \times 256 \times 256 = 16{,}777{,}216$ different colors! Indeed, that is almost
17 million different colors, a staggering number. Actually, some researchers believe that humans can detect up to 10 million colors and
intensities, so 17 million is reasonable.

IF YOU FIND 16,777,216 COLORS HARD TO BELIEVE AND IMAGINE, CONTINUE
READING HERE:
Why 16,777,216 different colors, and not 3×256? The 3×256 colors give you
256 intensities of red, and 256 of green and 256 of blue. However, they don't
give even a single combination of these three colors, so no yellows, no cyans,
no magentas, no golds, no browns, no yellows, *etc.*, and even no white and no
grays!

(Continued)

(Continued)

To understand the reason for the huge number 16,777,216, think of an analogy: three randomly chosen people and their birthdates. Each of these three people can have any of 365 birthdates (I ignore year of birth and leap years, and assume that each day of the year is equally probable for births). Now try to imagine how many combinations of birthdates are possible. The first person can have any one of 365 birthdates: for each of these 365 options, the second person can also have any one of 365 birthdates, making 365 × 365 = 133,225 possible combinations of the two. Now, for each of those 133,225 pairwise combinations, the third person can also have any of 365 birthdates, resulting in 365 × 365 × 365 = 365 × 133,225 = 48,627,125 three-way combinations: almost 49 million! Out of curiosity, you may now ask what is the chance that three randomly chosen people have a given birthdate, say 15 June (regardless of the year of birth): the chance for this to happen is about 1 in 49 million; for two people, like a couple, or a brother and sister, it is about 1 in 130,000.

Thinking of colors again, the same reasoning leads to a similar conclusion: the chance that three randomly selected colors from the RGB set are identical to a given color of given intensity is about 1 in 17 million. Similarly, the chance that three dresses have a particular color of the RGB set is about 1 in 17 million; for two dresses it is 1 in 256 × 256, or 1 in 65,536. The reason that I emphasize these huge numbers and tiny probabilities is to draw attention to the vast number of colors which we can generate: as it turns out, it is in fact very easy for you to generate those 16,777,216 colors with your computer, as I will show in Section 6.2.3.

4.4   Conversion from color to gray, and conversion from gray to color

Why would you want to convert a <u>***color***</u> ***picture to a*** <u>***gray-scale***</u> ***picture?*** Perhaps to create a classic old-time look? Or because you like the gray-scale style (often called **black-and-white** or **black/white** or **b/w**) in artistic photography? Or to publish a book more cheaply, since printing color is expensive? Nowadays, we rarely need or wish to change a color picture to gray-scale. But it is easily done at the touch of a button in most graphics software.

On the other hand, there was a time when television programs were broadcast in color while many viewers still only had a "black-and-white" television set: those television sets had to rapidly and automatically convert the incoming color images into gray-scale for immediate viewing. If you wish to know more about how this is done, please read the following note.

A MORE TECHNICAL NOTE: HOW ARE COLORS CONVERTED TO GRAY-SCALE? As we have seen (in Sections 4.1 and 4.3), in the RGB system any color can be decomposed into a red, a green and a blue component, with intensities ranging from 0 to 255. We have also seen that when the three color components are equally intense (equally bright), the resulting color is gray; this gray can range from white to black. So, if we even out the intensities of the red, green and blue components to become equal to each other by redistribution of intensity, we can obtain a gray of the same average intensity as the original color. The easiest way to do this redistribution of intensity is to use the average of the red, green and blue intensities. The simplest average is gray = (red + green + blue)/3: I call it "equal-weight average". For example, the basic yellow color is composed of maximum-intensity red (level 255), maximum-intensity green (level 255) and zero-intensity blue (level 0), which we can label (255,255,0); then the equal-weight gray conversion of this yellow is (255 + 255 + 0)/3 = 170. This average intensity 170 is then given to each of the three colors red, green and blue: we can label this (170,170,170). Since we then have equal intensity levels 170 of red, green and blue, the resulting color is indeed gray of intensity 170, which is roughly half of white's intensity of 255.

Of course, black, white and gray should not themselves change when converted to gray-scale: indeed, black remains (0 + 0 + 0)/3 = 0, white remains (255 + 255 + 255)/3 = 255, and gray of intensity 170 remains (170 + 170 + 170)/3 = 170.

This conversion to gray-scale is illustrated in the left and middle columns of Figure 4-11 for nine basic colors. For instance, as described above, yellow with intensities of red, green and blue at levels 255, 255 and 0, labeled there as (255,255,0), is redistributed to become (170,170,170), a relatively light gray.

We notice in Figure 4-11 that basic red, green and blue (the three topmost colors) are all converted to the same dark gray (85,85,85) using equal weights (middle column). However, to our eyes, the three basic red, green and blue do appear to be very different in brightness: green appears much brighter than red, which appears much brighter than blue, so green should ideally convert to a lighter gray than does blue. Do you agree? This can also be seen in the color triangle of Figure 2-5 and the color circles of Figure 2-13. More precisely,

(Continued)

(*Continued*)

Conversion from color to grayscale: <u>basic</u> colors.

These colors convert to these simple grays or these realistic grays

with compositions (red,green,blue)	using equal weights (0.33, 0.33, 0.33) for (red,green,blue)	using usual weights (0.299,0.587,0.114) for (red,green,blue)
red (255,0,0)	gray (85,85,85)	gray (76,76,76)
green (0,255,0)	gray (85,85,85)	gray (150,150,150)
blue (0,0,255)	gray (85,85,85)	gray (29,29,29)
yellow (255,255,0)	gray (170,170,170)	gray (226,226,226)
magenta (255,0,255)	gray (170,170,170)	gray (105,105,105)
cyan (0,255,255)	gray (170,170,170)	gray (179,179,179)
black (0,0,0)	black (0,0,0)	black (0,0,0)
gray (128,128,128)	gray (128,128,128)	gray (128,128,128)
white (255,255,255)	white (255,255,255)	white (255,255,255)

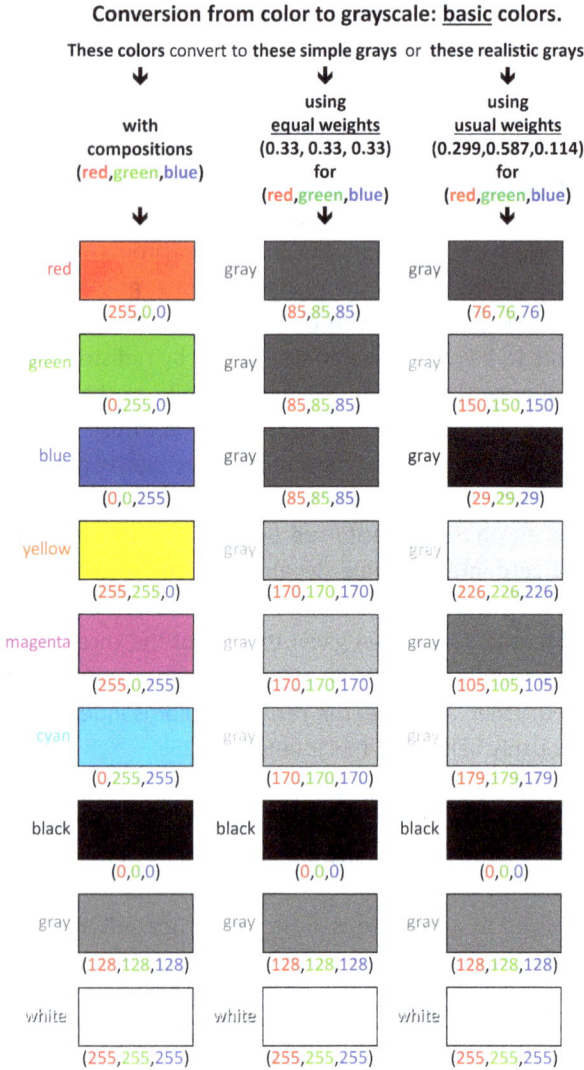

Figure 4-11: This graph shows how basic colors (left column of boxes) are transformed to gray using two different methods. The labels, such as (255,255,0) for yellow, give the decomposition into the basic colors red, green and blue (yellow combines red and green). The middle column of boxes uses equal weights in averaging the three red, green and blue components of each basic color. In the right column of boxes, the average uses unequal weights to account for the higher sensitivity of human eyes to green, and their lower sensitivity to blue; therefore, green converts to a lighter gray and blue converts to a darker gray.

Conversion from color to grayscale: <u>more complex</u> colors.

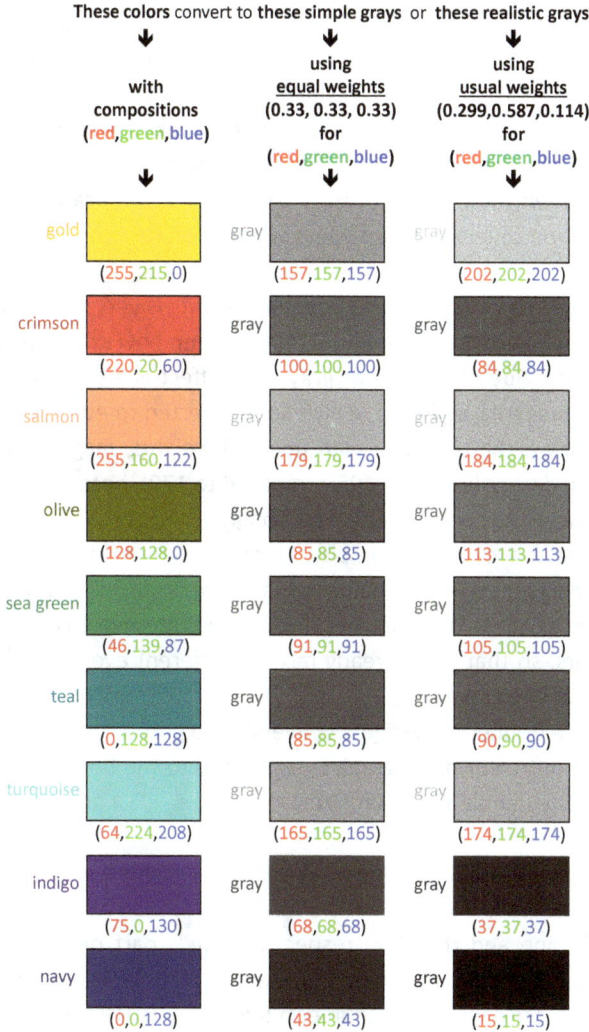

These colors convert to **these simple grays** or **these realistic grays**

	with compositions (red,green,blue)	using equal weights (0.33, 0.33, 0.33) for (red,green,blue)	using usual weights (0.299,0.587,0.114) for (red,green,blue)
gold	(255,215,0)	gray (157,157,157)	gray (202,202,202)
crimson	(220,20,60)	gray (100,100,100)	gray (84,84,84)
salmon	(255,160,122)	gray (179,179,179)	gray (184,184,184)
olive	(128,128,0)	gray (85,85,85)	gray (113,113,113)
sea green	(46,139,87)	gray (91,91,91)	gray (105,105,105)
teal	(0,128,128)	gray (85,85,85)	gray (90,90,90)
turquoise	(64,224,208)	gray (165,165,165)	gray (174,174,174)
indigo	(75,0,130)	gray (68,68,68)	gray (37,37,37)
navy	(0,0,128)	gray (43,43,43)	gray (15,15,15)

Figure 4-12: As Figure 4-11, but for conversion of more complex colors to gray.

it is estimated (by asking people, as was systematically done already over a century ago) that the basic green appears to be almost 2 times brighter than the basic red, while the basic red appears to be about 3 times brighter than the basic blue; this implies that the basic green is over 5 times brighter than the basic blue!

(Continued)

(Continued)

It therefore makes sense to take into account our eyes' very different sensitivities to basic red, green and blue when converting colors to gray. That is in fact the usual approach, which is illustrated in the right column of Figure 4-11. Here a somewhat different average is taken, with unequal weights, using the formula: gray = $0.299 \times$ red + $0.587 \times$ green + $0.114 \times$ blue for a color labeled (red, green, blue). This formula gives green significantly more weight (almost 59%), and blue significantly less weight (about 11%), compared to red (almost 30%).

The effect of this average with unequal weights is very visible for the basic colors shown in Figure 4-11: see in the right column how green becomes much lighter gray, and blue much darker gray. This effect is also significant for the colors yellow, magenta and cyan, which are converted to equal gray intensities with the equal-weights average: with unequal weights, magenta in particular becomes much darker (intensity 105 compared to 170), because our eyes don't detect its red and blue components so strongly.

The conversion of more complex colors is shown in Figure 4-12. Here we see less difference between the equal-weights and the unequal-weights averages; the reason is that most of these colors are more complex mixes of the basic colors, so that they already mix the different eye sensitivities before we convert them to gray-scale.

An example of unequal-weights conversion to gray is shown in Figure 4-13, where the top left color photograph is converted to the bottom left gray-scale image. Another example is shown later, in Figure 5-11, a gray-scale version of Figure 2-1.

A final point: the abovementioned unequal-weights formula for gray was developed specifically for the early PAL and NTSC television systems used in parts of Europe and the USA, respectively (it is part of a standard called ITU-R BT.601). A more recent formula (in standard ITU-R BT.709) was developed more recently for use with high-definition television (HDTV); it gives even more weight to green (namely gray = $0.2126 \times$ red + $0.7152 \times$ green + $0.0722 \times$ blue), and again affects the basic colors more than the more complex colors.

Why would you want to colorize a picture, namely convert a _gray-scale_ picture to a _color_ picture? Many old pictures and movies are gray-scale ("black-and-white"): obviously, we may prefer to see them with colors. In the days before affordable color photography, important portraits, wedding photographs, *etc.*, were often shot in gray-scale before color was added, mostly by hand. Similarly, color can be added to older gray-scale movies. This process is often called **hand-coloring** (mostly for single images) or **colorization** (mostly for movies).

Adding color is a very labor-intensive task, which also involves guessing the original colors (a gray-scale image cannot tell us the original colors of the scene that was photographed). This is especially time-consuming for movies, as each frame must be colorized individually and consistently. Computers are now helping accelerate the colorization process: software is evolving rapidly to increasingly automate the process.

In colorization, typically, a gray-scale image was painted over with thin transparent layers of colored paint that allowed the gray-scale image to be seen through a haze of color: it is like hanging a colored transparent plastic sheet in front of a gray-scale painting. This requires rather thin layers of paint to avoid overwhelming the underlying gray-scale image. As a result, colorized images are often only "weakly" or "softly" colored: hence the pastel coloring characteristic of colorized images.

A digital approach by computer can more realistically modify the gray-scale image itself to match the addition of color: basically, a computer can change the gray-scale image itself by incorporating the colors directly into the image, instead of relying on a "filter" covering the intact gray-scale image like paint. Software may automatically assign common colors to standard objects like skin, trees, water and sky. Nice examples of scenes in London in 1924 and England's Cornwall in 1916 that were "AI enhanced and colorized" are available at YouTube[1]: look especially at the predominantly brownish colors and how they change over time in the same scene; also watch the blues of sky and water. We see very few vivid colors red, green, blue, yellow, *etc.*: instead, the colors are rather soft, foggy and unsaturated. ("AI" stands for Artificial Intelligence, an automated computational technique. The AI enhancement "involves motion-stabilisation, speed correction, contrast, brightness and sharpness enhancement, noise reduction, dust and

---

[1] https://www.youtube.com/watch?v=6xLRXrJ-H6I, by the British Film Institute, and https://www.youtube.com/watch?v=9NeOACPBQsA, by Rick88888888.

speckle removal, and upscaling to HD", where HD means high definition or fine detail.)

I have simulated the hand-coloring approach with graphical software, as shown in Figure 4-13: I painstakingly produced an overlay (seen at bottom right of the figure) that is semi-transparent, similar to thin layers of paint or plastic. I had the benefit of the original photograph to choose my colors, but the color choice was actually not straight-forward: the appearance of a semi-transparent color is not the same as the appearance of the corresponding opaque color, so the degree of

From color to gray, and from gray to color (colorization)

Figure 4-13: My photograph at top left is shown in gray-scale at bottom left (as obtained by the first unequal-weights formula mentioned earlier in this Section). I colorized it (top right) using semi-transparent overlays shown at bottom right; the overlays were produced in Microsoft PowerPoint (the white parts here are transparent overlays that leave the corresponding parts of the gray-scale image untouched).

transparency was an additional choice to be made; for this I found the HSL scheme of Section 4.1 to be convenient (it is analogous to choosing the right thickness of the paint layer). On the other hand, if the original colors are not available, nobody will notice if your colorization is not accurate!

Another issue is how much detail should be colorized: the photograph of Figure 4-13 has a lot of detail, such as the windows and the large crowd of people, taking hours of work (even so, I only did justice to some of the windows and inserted just a few color patches in the crowd). A simple landscape with greenery, blue water and blue sky would be much easier to colorize.

4.5 What have we learned in this Chapter?

We have seen that we can in principle create an infinite number of colors by combining just three primary colors. Three systems for doing so are RGB (using red, green and blue as the three primary colors), CMY (using cyan, magenta and yellow as the three primary colors) and HSL (using hue, saturation and luminance to select the color, its gray content and its brightness, respectively).

RGB is an additive process, with which red, green and blue are added to black to compose the desired color (for example in computer monitors, television screens, and other color displays). By contrast, CMY is a subtractive process, with which cyan, magenta and yellow are removed from white light by filtering out unwanted colors (for example in printing and painting). HSL is a convenient and intuitive approach that focuses first on the hue from red through yellow, green, cyan, blue and magenta, and then on degree of colorfulness (saturation) *versus* colorlessness, and finally on intensity (luminance).

In practice, there is no need for an infinite number of colors, since our eyes cannot distinguish colors or intensity levels that are almost identical. Using 256 levels of color intensity, we can combine the three primary colors red, green and blue to generate 16,777,216, or almost 17 million, different RGB colors. All these colors can easily be produced within Microsoft Word or PowerPoint or Excel, for example, as shown in Section 6.2. However, the RGB color scheme cannot produce the solar spectral colors: we can only approximate the solar spectral colors with RGB colors.

When we speak of an object's color, we actually mean its color when seen in white light. More generally, however, a reflected color consists of the RGB components that are common to both the incoming light (which need not be white) and the surface.

Using RGB components of colors, it is easy to convert color images to gray-scale images ("black-and-white" images): this is readily done with existing software. The reverse process of converting gray-scale to color images is a bit more complicated, especially for movies, mainly because the colors to be introduced are unknown and should be consistent: software is becoming available to perform this conversion, but is far from perfect so far.

5

Recording Images

We now have a good grasp of how colors are related to each other and how we can describe and compose them in terms of basic colors like red, green, blue, cyan, magenta and yellow.

*We can now address the general question of **recording images: How can we make a good image of a scene that we observe with our eyes or that we photograph with a camera?** Let's start with our vision, which we will discuss in some detail as it teaches us much interesting physics. We will then consider photography with chemical films, and finally modern digital photography.*

—◦◦◦◦◦—

5.1 Human vision

Our eyes are amazingly powerful instruments that silently perform remarkable feats in combination with our **brain**. Let's explore how that happens.

5.1.1   *About cones and rods*

How do our eyes detect light and color? You probably know about the **cones** and **rods** in our eyes: these are the individual detectors of light and color (we will look at them closely in Section 5.1.2). Most people have three types of cones (they get their name from their pointed conical shape): one type of cone is more sensitive to red, another type is more sensitive to green and the third type is more sensitive to blue.

We also have rods (with a cylindrical rod-like shape) that are sensitive to light of any color, but especially to weak light: the rods are thus most useful in weak lighting, such as at night. The rods give a black-and-white view, better described as a gray-tone or gray-scale view.

Are there differences in color detection between people? Like our other senses (hearing, taste, smell, touch, motion, *etc.*), our **vision** varies from person to person. Some people are noticeably color-deficient or even color-blind, as we will discuss in more detail in Chapter 8. But even people with "normal" vision see the same object at least somewhat differently.

People with normal vision see color differently because of variations in the eyes' cones and rods. As we discussed earlier, we see color as a combination of red, green and blue: the exact amounts of red, green and blue determine the color (and intensity), for example in an RGB representation. But if your and my cones detect different relative amounts of red, green and blue coming from the same source of light, you and I will conclude different colors, even though they should be the same. We usually would not be aware of that difference of perception, because it is very difficult for us to compare what color you and I see: in fact, this difficulty of comparing is a serious issue also with color-deficient and color-blind people, to which we will return in Chapter 8.

However, we can compare colors of different objects and tell whether they look the same or different. Suppose you want to repaint a damaged object with its original color, but the "formula" (composition) of the original paint is not known. For example, you may want to repaint a damaged part of a wall in your home; or a repair shop may need to repaint a damaged panel of your car to match the other panels. By trial and error you or the repair shop may match the original color very well, as far as your eyes are concerned. Nevertheless, someone else may see a mismatch between the new and original paints, because the new paint likely is chemically different and indeed can appear visually different to his or her eyes.

We can conclude: Different people may see the same light color differently. And two light sources (whether paint, ink or electronic display) may appear to have the same color to one person but two different colors to another person.

5.1.2 *Sharpness of view: Visual acuity*

The following question may initially appear rather narrow, but it will open the door to very interesting aspects of our vision.

How sharp is your vision <u>away from</u> the center of your eyes' view? Do this simple experiment: focus and fix your view on a single word in the center of a page which is full of text, such as this page, with normal page size and letter size, and a normal, comfortable eye-to-page distance. Now, <u>without moving your eyes</u>, try to read words away from your focus point: can you read all the way to the left and right edges of the text without looking there? Probably not! You probably are not even able to read text that is several words away from your focus! To read such text, we instinctively turn our eyes toward it; that of course is why we sweep our eyes across the page as we read line by line! Wouldn't it be more convenient if we could read without constantly sweeping our eyes to the left and the right?

We can ask a very similar question about watching people's faces in a crowd: **How many faces can we distinguish at once without turning our eyes?** Clearly, we can only see details of one face at a time: we have to turn our eyes to examine and distinguish other faces.

This shows: We read and scrutinize primarily with one small part in the center of the field of view of our eyes.

Our full **field of view** is defined by the **retina**, which contains the cones and rods in the back of our eyeballs. The retina allows us to see all the way to the left and right, about 90 degrees from the forward direction, without turning our eyes, while our view is more limited up and down, by our eyebrows and cheeks.

The center of the retina is called the **fovea**: it gives us the sharpest part of our field of view and is about 0.3 millimeter (1/100 inch) across, similar to the cross-section of very thick human hair. The fovea's size translates to an angle of sharp viewing outside the eye of about 2 degrees only: this corresponds roughly to the apparent size of the nail of our little finger on our outstretched arm; thus, at a reading distance of about 30 centimeters (about 1 foot), 2 degrees corresponds to about 1 centimeter (0.4 inch) on a page, which covers only about one word. The fovea is the only part of the eye's view that is sharp enough for normal reading. Away from the fovea, the density of cones decreases; therefore, to read other text, we must turn our eyes so our fovea can focus on it.

Similarly, we must turn our eyes to examine different faces. To be more quantitative, the 2-degree angle of sharp viewing spreads to about 20 centimeters (8 inches) at a distance of 5 to 6 meters (15 to 20 feet), as shown in Figure 5-1: 20 centimeters is just about the size of a face. Therefore, we can see details of only one face at a time at a distance that is typical for a face across a room or in a crowd; from half a meter away, or about 2 feet, we can scrutinize only a small part of the face at a time: only one eye or the mouth or the nose!

Size of scene imaged onto fovea of human retina

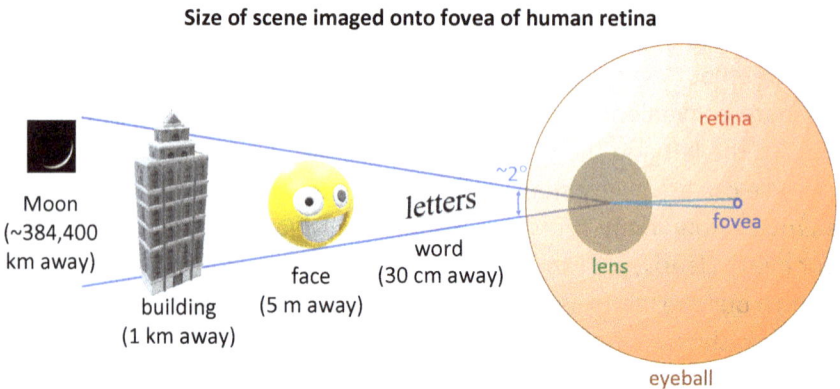

Figure 5-1: This schematic shows how far away you can see details of typical objects, as they are imaged onto the human eye's fovea. Here we see the eyeball from the back, looking forward toward the word "*letters*" at 30 centimeters, a face at 5 meters, a building at 1 kilometer or the Moon almost 400,000 km away (the Moon spans about 0.5 degrees, as does the Sun, but don't look at the Sun!). The blue cone at left in reality has an opening angle of about 2 degrees, which is exaggerated in the perspective view used here. The fovea is about 0.3 millimeter across (about 0.01 inch), which is one hundredth of the eyeball size of about 25 millimeters (2.5 centimeters or 1 inch).

The 2-degree angle also allows us to examine only one building of size 30 meters (100 feet) at a distance of 1 kilometer (0.6 mile). For comparison, the Moon seen from Earth is only half a degree across, so four Moons aligned against each other would fit our fovea.

Now try the following: increase the distance from your eyes to the text you are reading and estimate the distance where you lose the ability to read it. Is it about 1 meter (three times the normal reading distance, or 3 feet, or one arm's length), or is it twice that, or even more? If you find that the text becomes too small to read at a distance of 1 meter, then the 2-degree angle of sharp viewing covers about 3 centimeters (about 1 inch) or 3 words of text across the page. If you can read at a distance of 2 meters (about 6 feet or the length of your body), then you can probably read about 6 words without turning your eyes.

Next ask yourself: *How sharp is your vision close to the center of your view?* **Sharpness of vision** is also called **visual acuity** or **resolving power**. It describes how small are the details that you can distinguish in a scene. Normal visual acuity is often called **6/6 vision** or **20/20 vision**, while a larger second number, such as 6/9 or 20/30, means visual acuity that is below normal; a smaller second number, such as 6/4 or 20/15, means visual acuity that is better than normal.

WHAT EXACTLY IS 6/x OR 20/y VISION? First, "6/6 vision" means normal vision at a distance of 6 meters, and "20/20 vision" means normal vision at a distance of 20 feet (which is close to 6 meters); in other words, 6/6 or 20/20 vision is average for healthy human eyes. Note that this measure of visual acuity does not say anything about vision nearer or farther than 6 meters or 20 feet; it also does not mean "excellent" or "better than normal" vision.

Second, if you have "6/9" vision, then another person with normal vision can see from 9 meters what you see from 6 meters; in other words: what another person sees well from 9 meters away, you can see well only from 6 meters away. Similarly, if you have "20/30" vision, then another person with normal vision can see from 30 feet what you see from 20 feet; in other words: what another person sees well from 30 feet away, you can see well only from 20 feet away. Let's explain that.

Normal visual acuity allows you to distinguish two small dots that are separated by an angle of about 1 arc minute (1/60 of a degree). For example:

(Continued)

(Continued)

that angle corresponds to a separation of about 3 centimeters (~1 inch) between small lamps placed at a distance of 100 meters (~330 feet) from you; or you can distinguish two dots that are separated by about 0.2 centimeter (2 millimeters or ~0.08 inch) and located 6 meters (~20 feet) away; in text or images at a reading distance of 30 centimeters, you can distinguish dots separated by about 0.01 centimeter (0.1 millimeter or ~0.004 inch). For comparison, using the normal font size 12 on paper, the dot size on the letter i in "Times New Roman" is about 0.03 centimeter (0.3 millimeter or ~0.01 inch), which is three times the distance you can normally distinguish at 30 centimeters.

Acuity is calculated as 1 divided by the separation between those dots, as measured in arc minutes: so "normal" vision is 1/(1 arc minute) = 1 = 6/6 = 20/20. If you have 6/12 or 20/40 vision, your visual acuity is 6/12 = 20/40 = 1/2, implying that two dots must be separated by at least 2 arc minutes for you to distinguish them.

What limits our visual acuity? Can you think of reasons that we cannot see smaller details than we discussed above? The two main reasons are: the size of our cones, and poor focusing of light onto our retina.

Let us first think about the **cones** which detect light in our retina: Our visual acuity, and thus our sharpness of view, is limited first by the size of our eyes' cones. (We ignore the rods in this discussion, because they are largely absent from the fovea and therefore contribute little to visual acuity. Indeed, in weak light our vision is not as sharp as in strong light.)

How much detail can our cones detect? Here we are basically asking how close together the images of two points can be on our retina, while we can still tell them apart as separate points. Imagine looking at two very sharp points of light, perhaps two distant lamps or two stars in the night sky; you can also look from a distance at two tiny white dots on a black background, such as this pair of dots: ▪▪

Our eye's **lens** will image those two points of light onto the fovea: if the two points of light are so close together that they are imaged on the same cone, then we cannot distinguish them; if they are imaged on two adjoining cones, we see the two points as a single slightly larger source of light because there is no cone between the two points to tell us of a gap of light between them; to see them as separate points of light, there should be at least one cone that is <u>not</u> illuminated between the two cones that are illuminated. This situation limits our visual acuity, as we already saw above: Visual acuity measures how close together points of light or details in a scene can be distinguished.

So, we must next ask: *How close together are the cones in our fovea?* Cones in the human fovea are about 0.5 micrometer across: 0.5 micrometer is therefore also how close the cones can get together. (A **micrometer** is a thousandth of a **millimeter**, which in turn is a thousandth of a meter; 1 micrometer is about four hundred-thousandths of an inch or 0.04 **mil**.)

As you can well imagine, a micrometer is too small for us to see, since this distance would be imaged onto an even smaller area of the fovea than the cones themselves. Therefore, we need a microscope to see individual cones in the retina.

It is interesting to note that the individual cone size increases from about 0.5 micrometer near the fovea to about 4 micrometers farther away, approximately proportionally to the distance from the fovea. This is one reason why your eyes see progressively less detail away from the center of your field of view: at the edges of your field of view you see at least 8 times less detail than in the center. Another reason for such loss of detail is that the retina is approximately spherical and not planar: as we know from cameras, optical lenses focus a scene on a planar film or detector, not on a spherical detector like the retina; therefore, image details farther away from the fovea become gradually more blurred (out of focus).

Now we are in a position to connect the size of the cones in our eyes to the size of the text that we can read (we will discuss the role of the lens focusing in Section 5.1.3): *How many cones do we use to read the smallest readable text?* This question basically asks how many cones exist within the fovea. The dimensions given above for cone and fovea imply that there are around 600 cones across the fovea, counting along a straight line. Since the fovea is roughly circular, the fovea then includes close to 300,000 cones within its disk, now counting in two dimensions (for comparison, the much larger retina counts about 6 million cones overall).

If we count along a horizontal line, the smallest readable line of text is thus imaged onto about 600 cones across the fovea. How many words of text then fit on the fovea?

We reasoned before that we need three cones to distinguish two light points (only when the middle cone is not illuminated can we tell that there are two separate points of light). Similarly, we can say that, for example, the letter "m" requires at least seven cones to be recognized: one cone for each of the three legs of the letter, and one cone for each of the white spaces between and around those legs (other cones will be needed for the vertical dimension of the letter). Figure 5-2 illustrates this, but it is very difficult to distinguish the letter "m" in this image, even if you look at it from a distance!

It is clearly better to double that number of cones to, say, 15 cones, for good recognition of a letter, as shown for the same letter "m" in Figure 5-3.

Figure 5-2: Simplified schematic of cones in the human fovea responding to viewing a black letter "m" on a white background. The circles represent "red", "green" and "blue" cones: their color intensity corresponds to the light intensity falling on them; full intensities correspond to white illumination. The right-hand image shows the outline of the letter "m" overlaid on the left-hand image; cones overlapping the black letter are proportionally darker.

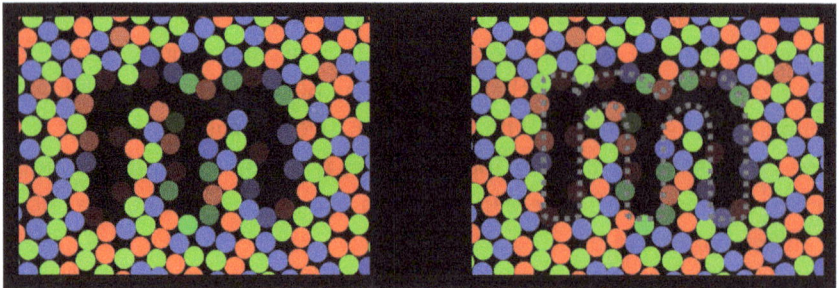

Figure 5-3: As Figure 5-2, but with cones drawn at half size and half separation to allow better detection of the letter "m".

Then, 40 letters of width 15 cones will fit in the 600 cones of the fovea (along one line): that is about 6 words. However, 6 words is somewhat more than the one or two readable words that we deduced above by looking at text with our eyes. We can therefore conclude: **Our eyesight is not as sharp as our eye cone size would imply.** We will therefore discuss the role of the lens in Section 5.1.3.

HERE IS AN INTRIGUING PUZZLE: Looking at the random layout of cones in the retina, as in Figure 5-3, one may ask another question: **How can we distinguish a straight line from a curved line in a scene?** Look at Figure 5-4: you should be able to tell which two lines are straight, without glancing along the page. The puzzle is that you can do this even though there is no straight alignment of cones in the retina that may serve as a guide to identify a straight line in a scene.

Figure 5-4: Which of these lines are straight?

There is another interesting related observation: if you wear glasses, have you noticed that, when you change the prescription of your glasses, lines that looked straight before may suddenly look curved with the new glasses? Intriguingly, within a few days those lines look straight again: it is as if your brain corrects the interpretation of lines in such a way that lines that you know are straight will indeed appear straight!

Similarly, how does our brain identify a letter? Let's consider the letter "m" in Figure 5-2 or 5-3. The cones in the retina are not pre-arranged as a letter "m" (or any other letter or character). Another complication is that each cone leads to the brain through long nerves: how does the brain know which cone lies next to which other cones, so as to identify patterns like letters?

These fascinating questions open up a whole new subject: how does the brain interpret the signals sent by the cones (and rods) in the retina? It is known, for example, that the optical nerves form a neural network that first identifies simple line segments, like the signs "—" or "/" or "\" or "|". In a next stage, slightly more complex shapes are identified, such as crosses "+", "x" or circles " ○ ", followed by even more complex shapes, including letters,

(Continued)

(Continued)

etc. The identification of simple line segments and shapes is believed to occur in the optical nerve connections in or near the retina, where their mutual locations are still clear. This analysis of images is a fascinating topic that we can only address briefly in this book: see Section 5.1.5. We will also encounter the brain's effect again in Chapter 10 when we discuss optical illusions: these illusions provide useful insight into the role of the brain in our vision. Indeed, the brain plays a major role in vision.

5.1.3 *Focusing in the eye*

We now turn to the second major cause for a lack of sharpness in images formed by the eye: imperfect **focusing**. *What causes imperfect focusing?* The short answer is that our eyes' **lenses** and our eyes' shapes are not optically perfect. These and other imperfections degrade the sharp focus that we need for sharp viewing; such imperfections also worsen with age, further reducing the achievable sharpness.

You are probably familiar with the basic functioning of the human eye, as illustrated and explained in Figure 5-5. You will find much more useful information about vision and eyes at the Hyperphysics website.[1]

BUT YOU MAY NOT REALIZE ONE SURPRISING FACT: The eye's lens does only about 20% of the work of focusing light onto our retina. About 80% of the focusing, namely most of the bending of light rays, is actually done by the interface between the cornea and the air. We will discuss how a lens focuses light in a note a bit further below.

The reason why the cornea (the outer curved front of the eye shown in Figure 5-5) plays this larger role is that its interface with the air causes a relatively large bending of light rays (similar to the water-air interface). By contrast, the interfaces between the lens and the surrounding liquids of the eye cause less bending of light. We see the same effect with a film of oil on water: the air-oil interface bends light more than the oil-water interfaces because the oil and water are optically very similar to each other, while both are optically quite different from the air.

---

[1] http://hyperphysics.phy-astr.gsu.edu/hbase/index.html and in particular:
http://hyperphysics.phy-astr.gsu.edu/hbase/vision/imagformcon.html.

Focusing in the eye

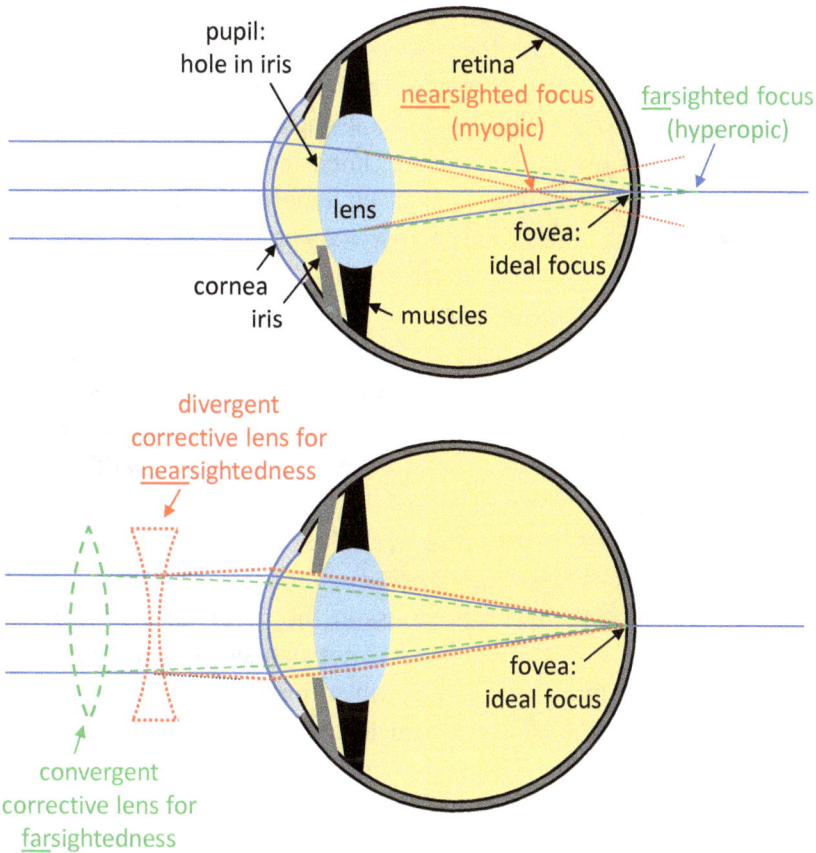

Figure 5-5: These two sketches show the relaxed human eye watching a distant source of light. The lens adjusts the focusing for viewing less distant objects by tensing its muscles. The top sketch shows parallel light rays (blue) entering the unaided eye from the distant object. Ideally the light is focused on the fovea, as shown by the converging blue rays. With farsightedness (dashed green rays), the light focus falls behind the fovea: so the light is spread out on the retina, forming a blurred image. Conversely, with nearsightedness (dotted red rays), the light focus falls in front of the fovea: it is also spread out on the retina, again forming a blurred image. The bottom sketch shows how glasses (spectacles) can correct these imperfections. For farsightedness (green), glasses with convergent lenses (which are thickest in the center) move the focus of the light forward onto the fovea. Analogously, for nearsightedness, glasses with divergent lenses (which are thinnest in the center) move the focus of the light backward onto the fovea.

The main purpose of the eye's lens is to adjust the eye's focusing to view objects at different distances. The lens does this by adjusting its shape, namely by changing the curvature of its outer and inner surfaces, which is accomplished with the help of muscles surrounding the lens.

The exact shapes of the eye's **cornea** and **lens**, as well as their distances from the fovea, are crucial to achieving a sharp focus and thereby a sharp image of everything we see. Imperfect shapes prevent sharp focusing, as in **farsightedness** (also called **hyperopia** or **hypermetropia**), and **nearsightedness** (also called **myopia**); both cases are illustrated in the top part of Figure 5-5. Such imperfections cause blurring of everything we see: the lens may not be able to compensate for this blurring, especially as the lens loses flexibility with age. Fortunately, corrective glasses placed outside the eye can normally compensate for these deficiencies: the bottom part of Figure 5-5 shows how **corrective glasses** can correct farsightedness and nearsightedness by adding to or removing from the focusing power of the eye's lens.

Another imperfection of the eye is **astigmatism**, which is more complex: it combines farsightedness in one direction, for example in the horizontal plane, with nearsightedness in the perpendicular direction, such as vertically. In this case, there is not one single focus point, but there are at least two focus points, which may not fall on the retina; this is compensated with **cylindrical** lens shapes instead of spherical shapes (a cylindrical lens is curved in one direction but not in the other direction, like a straight glass or bottle, see Figure 5-7).

As we will discuss in the following more detailed note about optics, the eye is not able to perfectly focus light. Images on the eye's retina therefore cannot be perfectly sharp.

What is the famous red-eye effect? The familiar **red-eye** effect is especially visible at night or with a camera's flashlight in photographs: a person's or an animal's eye lights up with a bright red color (many animals' eyes are built similarly to human eyes).

The red-eye effect is due to light focusing within the eye, both on the way in and on the way out. When a sharp beam of light enters an eye (of a human or animal), the light is focused on a small spot on the retina, as shown by blue arrows in Figure 5-6. Part of that light is reflected in all directions (red arrows); the reflected light is mostly red due to the blood coloring the retina. One part of the reflected light

Red-eye effect

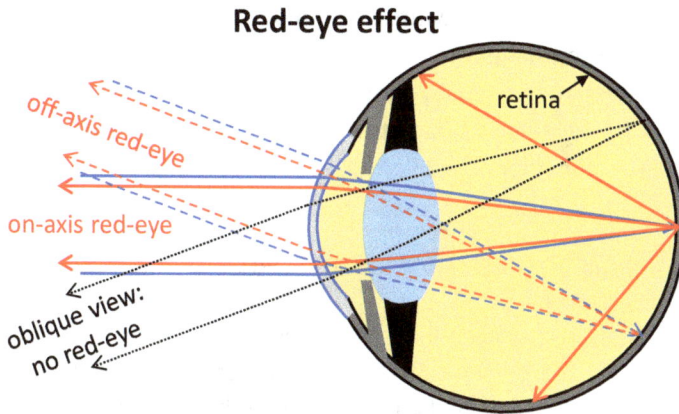

Figure 5-6: The famous red-eye effect occurs when light enters the eye (blue arrows) and is reflected by the retina; the reflected rays are focused by the cornea and lens in the direction (red arrows) which is opposite to where the light came from. This happens both on the optical axis of the eye (full lines) and off that axis (dashed lines). To avoid seeing or photographing the red-eye, one must view or photograph away from the incoming light beam (dotted black lines).

escapes through the **pupil** (the hole in the **iris**) and is focused back toward the original source of the light by the lens and cornea. Indeed, light follows exactly the same path in both directions through lenses; in this case the light is sharply beamed back toward its original source.

So if you look or photograph close to the incoming light beam (such as that from a torch light or camera flashlight), you will see the red light coming straight back at you from the eye; this can happen even in daytime or in a lit room if you use a powerful flashlight!

You cannot avoid the red-eye effect by asking the subject to look away from you, or by yourself moving off to the side with the light source: as the dashed lines in Figure 5-6 show, the red-eye effect still exists off the optical axis of the eye.

There is a relatively easy way to avoid seeing the red-eye: look or photograph away from the incoming light beam, for instance the oblique view (black lines) shown in Figure 5-6. You then look at a point on the retina which receives very little light from the incoming light beam. So avoid being close to the incoming light beam. This means keeping the flashlight of a camera well away from the camera lens; that is certainly not practical in a compact camera or smartphone camera! Therefore, modern digital camera software attempts to detect and remove the red-eye effect in the image after it is recorded.

We see that the red-eye effect is due to the reflected light from the retina being beamed back toward the source of light, for example a torch light or camera flashlight.

HERE IS A MORE DETAILED NOTE ON THE OPTICS OF THE EYE: HOW DOES A LENS FOCUS LIGHT?

We have discussed the need for a perfect eye shape to achieve a sharp focus and therefore a sharp image. So what is that perfect shape? To address this question, we must understand a bit more about how light is bent by a lens or any other transparent substance.

Let's look at my photograph of a glass full of water in Figure 5-7. The glass acts as a **cylindrical lens**: sunlight coming from the left is "focused" on the paper at right. (The cylindrical shape of the glass is convenient because its focus is really a vertical line along the glass, and not just one point as with a magnifying glass, for example: this way the "focus" can be imaged more completely across the paper thanks to the inclined sunrays.)

Now look at the bottom part of Figure 5-7. You will recognize there the same "focus" obtained by raytracing the light across the glass. (A spherical glass ball would give the same "focus" effect, but more intense and less easily photographed.)

You may object that our eyes' lenses are not cylindrical or even spherical. However, our reasoning will not be affected by that difference: the same focusing effects and defects appear in our real eyes.

Here, I intentionally use quotation marks around "focus" because Figure 5-7 does not exhibit the perfect convergence of rays that you would expect for perfect focusing: even though we can see there a very narrow spike of light, we also see many light rays that miss this "focus" point (as shown by the raytracing and the photographed light above and below the "focus" point). Much of the light misses the "focus" point, so the "focus" is actually spread out in a complicated way and will not be a sharp intense focus. (This is also a cause of "flaring" of intense light sources, which we discussed in connection with laser light: see, for example, Figures 2-10 and 2-15.)

One way in which the eye improves its focusing is by restricting the light rays that form the image. Notice how in the raytracing of Figure 5-7 the rays (drawn in green) that come in near the center line form a relatively sharp focus, while the rays that come farther off-center (drawn in red) increasingly miss the focus. The restriction to central rays can be achieved by squinting, namely by

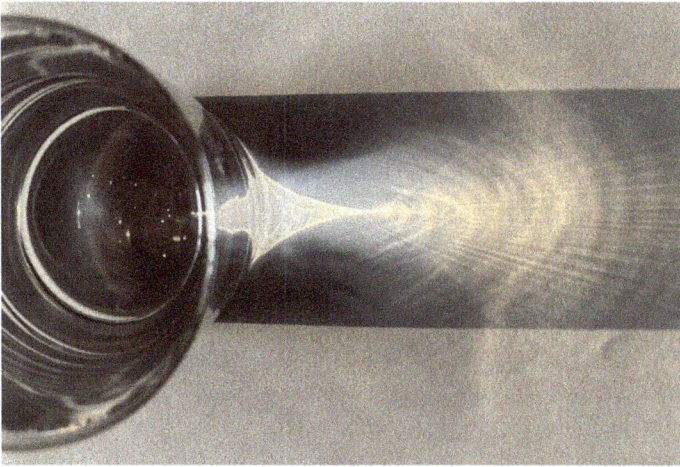

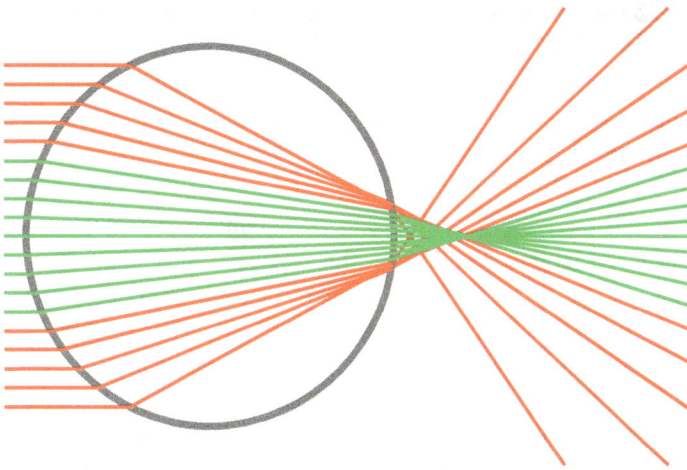

Figure 5-7: The upper image is a photograph looking down along the side of a cylindrical glass full of water (at left) standing on a horizontal white sheet of paper with sunlight coming in from the left (the wide gray band is the glass' shadow on the paper). Rays of sunlight converge on the paper just to the right of the glass, forming a spike with a pointed "focus". To the right of the spike, rays diverge. (The bright rays and arcs of light to the right are due to imperfections in the glass shape.) The lower image sketches the rays as they are focused by the cylindrical glass. The exact shapes and locations of the spike and "focus" point depend on the optical nature of the glass and water. The two intense curved lines flanking the spike and converging toward the "focus" point are called **caustics** in optics: they are locations where many light rays coincide; they are very clearly seen in the upper image of the spike.

(Continued)

(Continued)

narrowing the slit between our eyelids, at least in the vertical direction, as many people do if they have no corrective glasses. Another approach is that of the eye's iris: the smaller the pupil (the hole in the center of the iris), the better the focus will be. However, the restriction to central rays comes at a significant cost: less light will be accepted, making the image dimmer; conversely, since our pupil expands in weak light, our eye focuses less sharply in weak light.

We see that the circular profile of a cylindrical lens is not perfect for focusing light. The same holds true of a spherical lens. And it remains true with thinner sections of spheres, as used in corrective glasses (Figure 5-5) and in most optical instruments (such as cameras, magnifying glasses, microscopes and telescopes). The ideal non-spherical shape of a lens is mathematically very complex and also difficult to manufacture. Therefore, most manufactured lenses are based on spherical surfaces. It is also not surprising that human eyes are not perfect lenses, since nothing in nature provides a "blueprint" or construction mechanism to achieve the perfect shape; more surprising is actually how well most human eyes can actually focus light, especially with the help of corrective glasses.

ANOTHER EYE DEFICIENCY — CATARACTS: A common eye problem, particularly at older age, is cataracts, which affect the eye's lens and can develop gradually over many months. Cataracts can combine several imperfections that may occur simultaneously or at different times in both eyes. Clouding of the eyes is usually the most disturbing aspect, which is like looking through a dense fog, and leads to surgery to replace the lens. The eye's lens is normally colored slightly yellow, by absorbing some blue light, which the brain remarkably can compensate for by boosting the color blue, so white still looks white; in cataract surgery a new clear lens is inserted, giving a blue tint (and greater brightness) to everything, until the eye compensates back to make white look white again. Clouding may also be preceded by nearsightedness, blurred vision, double vision in each eye, halos around lights, or some combination of these.

At the time of writing I had cataracts that combined increasing near-sightedness and halos, but no noticeable clouding. This required occasional changes of glasses and caused amazing and spectacular sights at night: without glasses, each distant light was turned into a magical ball of fireworks, frozen in time. I had fun trying to explain what I saw, by asking many questions and doing a variety of simple experiments, since I could not find a direct illustration or explanation on the web. One difficulty I faced was that I could not take a photograph of what I saw, because the image was on my retina.

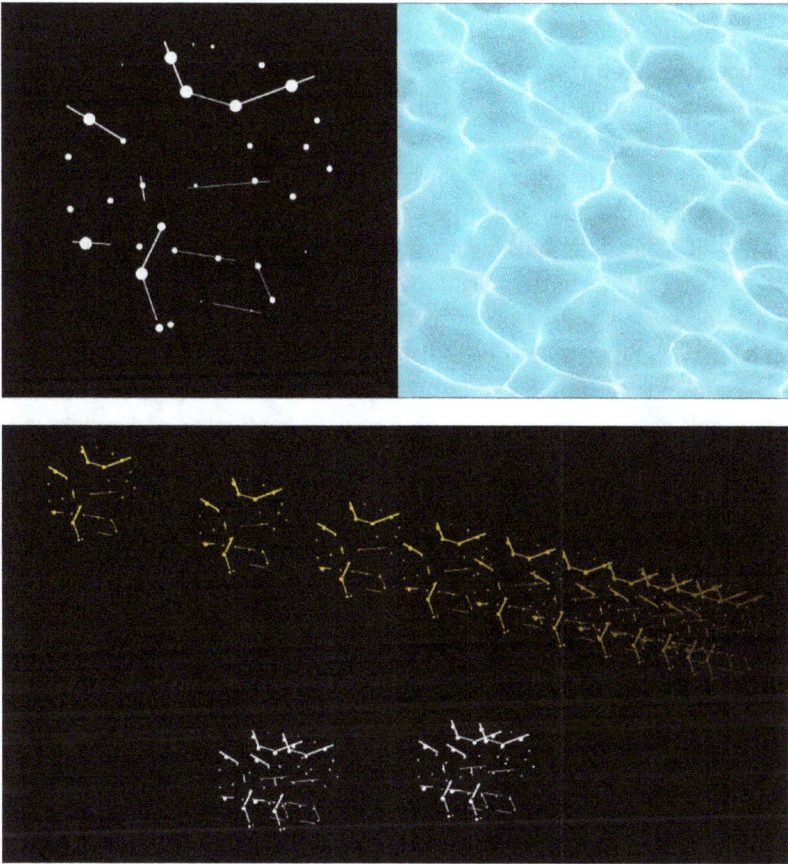

Figure 5-8: At top left is my simplified sketch of what my right eye saw when it looked at a single distant light without glasses, due to cataracts; with a perfect eye I should have seen a single light spot in the center. The actual image in my brain was more beautiful, more intense and shinier than I can draw it. The points and lines caused by my cataract resemble light effects due to the Sun in a swimming pool with waves, shown in the photograph at top right. The lower sketch shows a street scene with two cars below and a string of yellow-colored streetlamps above, as seen with my right eye: each light source is replaced by the same pattern (here reduced in size relative to the top left diagram).

I show at top left in Figure 5-8 a simple drawing of what I saw when looking at a single point of light, such as a strong distant lamp, at night without glasses. I saw tens of sharp brilliant points of light, some of which were interconnected by weaker streaks of light. This sharpness is remarkable considering that I was very nearsighted at that stage (without glasses, my distant vision was

(Continued)

(Continued)

very poor, while at a short range of about 15 centimeters or 6 inches, I had excellent vision, being able to even distinguish the smallest picture elements on a computer display, although not on a smartphone).

The light points and streaks which I saw all fit within a roughly circular disk, which looked about as large as the nail of my little finger on my outstretched arm (which happens to be similar to the angular span of the fovea). I believe that the size of that disk of lights was determined by my pupil size, since contracting my pupil (by shining more bright light into my eyes) shrank the disk by shaving off an outer ring of light points and streaks. Similarly, squinting my eyes or looking through a narrow gap between my fingers removed large parts of this disk, leaving only a narrow band of light points and streaks in the direction of the gap between my eyelids or fingers.

In Figure 5-8 I show only the view of my right eye. My left eye saw a similar pattern: also a disk of similar size with similar brilliant light points and streaks, but arranged quite differently within the same size disk. When opening both eyes the two patterns were simply superposed in my brain, doubling the number of points and streaks.

This disk of brilliant lights was identically repeated for each light source I looked at: it had exactly the same angular size and the same pattern of points and streaks, regardless of the light source's distance; but its color and brightness varied with those of the light source I was looking at, as sketched in the bottom image of Figure 5-8. Furthermore, the pattern of points and streaks rotated when I tilted my head sideways, thus following the orientation of my eye.

This pattern of points and streaks reminded me of another pattern that we are all familiar with: the pattern formed by sunlight falling onto the bottom of a swimming pool which has mild waves on its surface, as illustrated at top right in Figure 5-8. Here we also see relatively sharp points of light connected by weaker streaks of light (but the pattern in a swimming pool changes constantly due to the wave motion). What is happening in the swimming pool is that the curved top of a wave acts like a convergent lens (very much like the convergent lens of our eyes and the lens used in our glasses for farsightedness, see Figure 5-5), so it directs more light to some places than others; conversely, the curved bottom of a wave acts like a divergent lens, which spreads light out toward the brighter directions. I believe that this bending of light rays is the same effect that happens in my eyes. (Note that the pattern we see in a swimming pool is "projected" on the bottom of the pool; for us to see that pattern, it still needs to come out through the wavy surface on top of the pool and will therefore be further distorted upon exiting through the water waves. This does not happen with cataracts.) In optics such concentration of light by converging rays is called a **caustic** (this use of the word caustic is related to its other meaning of ability to burn and corrode due to high concentration).

By analogy with the Sun in the swimming pool, I conclude that my eyes also have a "wavy" disturbance. It does not have to be a wavy surface as in the pool, but it is more likely a variation in density in my eyes' lenses, which causes similar bending of light rays. The standard explanation for cataracts, including clouding, is that the substance in the lenses clumps up into small balls that act like little lenses and deviate light slightly, just like the waves in the swimming pool. In my case, those clumps or lenses happen to focus light on the retina sharply as dots or less sharply as streaks.

The bottom image in Figure 5-8 represents a night scene of two cars under a row of streetlamps, as seen by my right eye without glasses and with cataract. Each single lamp is replaced by the same brilliant fireworks ball, except for its color and intensity, which are the color of the individual lamp and its apparent weakening due to distance. Also opening my left eye adds more light points and streaks to each fireworks ball. One remarkable aspect is that the ball for each streetlamp and each car lamp has the same apparent size, no matter how close or far it is.

YOU MAY STILL WONDER WHAT PRECISELY CAUSES THE BENDING OF LIGHT RAYS. For example, why does a light ray bend upon entering and exiting a glass of water, or upon entering our eye's cornea and while passing through our eye's lens? This process, which is called **refraction**, is due to the **wave** character of light combined with its different speed of propagation in air *versus* glass or water. Refraction cannot be understood otherwise. Other types of waves, such as water waves and sound waves, also can undergo refraction, even though these waves have quite different physical origins. It is important to realize that we must think in terms of waves to understand the bending of light in lenses. We will therefore postpone this very interesting topic to another book devoted to the highly important subject of waves.

5.1.4 *Animal vision*

Let's turn to another topic related to human vision. You probably have asked yourself: ***How does animal vision compare with human vision?*** In this regard, it is of interest to learn that human vision itself evolved over time. Indeed, it has been discovered that early humans only had two kinds of cones in their eyes: "red" and blue. Only later did the earlier "red" cones split into today's red and green cones to give three-color vision. This early two-color vision of humans may be related to some forms of present-day color deficiency and color blindness: as we will discuss in some detail in Chapter 8, some color-deficient and color-blind people cannot easily distinguish red from green. (Note that here we use the phrases "two-color" and "three-color" vision in the sense of having two or three types of cones detecting two or three basic colors, such as red and blue, without counting their many combinations or intensities.)

Since animals are related to humans in different degrees, it is interesting to compare their and our visual systems. Such comparisons open up fascinating evolutionary questions, but we will not attempt to answers these here.

We can start with the question: ***How many colors do animals detect?*** Perhaps surprisingly, many animals actually detect four basic colors (and their combinations)! These four colors include essentially the same three colors that humans detect, with some variations. However, additionally, many animals detect **ultraviolet (UV) light**; as we discussed in Section 2.5, ultraviolet is invisible to the human eye, although it exists in the solar spectrum (ultraviolet lies to the right of blue in Figure 2-11). These animals include a variety of dogs, reindeer, fish, birds, insects, and spiders. You probably know that many insects are attracted to **black light**, which is ultraviolet light that spills over slightly into the visible blue/violet end of the solar spectrum (it is called black light because most of it is ultraviolet light that is invisible to us): it is used to attract insects to "bug zappers" that electrocute them. We have no idea how these animals perceive the ultraviolet color, but it must give a wider range of hues than humans are familiar with.

On the other hand, most mammals see only two colors: many dogs, cats and horses see yellow and blue (and their combinations), but not red. This is similar to what many color-deficient people see (as further discussed in Chapter 8).

An astonishing case is that of the mantis shrimp: seemingly it detects as many as 12 to 16 colors, but it is not clear how it uses all those colors! By contrast, there are animals that see only one color:

Figure 5-9: As Figure 2-1 but without the color red, leaving green and blue.

Figure 5-10: As Figure 2-1 but without the colors red and blue, leaving only green.

Figure 5-11: As Figure 2-1 but in gray-scale.

these include cows (which see red-to-orange) and sharks. It is difficult to tell whether they perceive that single color as we see gray or as something more cheerful.

So we may also ask: how useful is it to be sensitive to more colors? (We will encounter this situation again with color blindness in Chapter 8.) We illustrate the effect of removing one color in Figure 5-9, where red is removed from Figure 2-1, leaving green and blue. In Figure 5-10 we remove both red and blue, leaving only green. And for comparison, in Figure 5-11, we replace all three colors by gray (which is a realistic average over the three colors, see Section 4.4). It is obvious that with fewer colors, it is harder to distinguish different flowers (however, we can of course still distinguish flowers through their different shapes and sizes).

Another question is: **How many eyes or light sensors do animals have?** Most familiar animals have two eyes like we do. In some animals (rats, chameleons), the two eyes can turn independently of each other, giving two separate views simultaneously. By contrast, the two views from our two eyes are combined into a single view in the brain; we also use differences in perspective from our two eyes to gauge the distance to objects that we see (this **stereoscopic** vision relies on our eyes squinting closer together or "crossing" to see closer objects).

But some animals have more eyes, such as the eight eyes surrounding the head of the jumping spider. Others have even more eyes: clams have a ring of simple light sensors that tell them when to close the two clam shells; many insects have hundreds or even thousands of tiny eyes that together produce an integrated view of their surroundings.

A panoramic view of the surroundings is common also with many animals that have only two eyes looking to the sides of the head, such as cows, horses, fish and birds. These see all around except in some cases directly backward, because their bodies obstruct that view, unless they turn their heads.

Now I invite you to ponder the question: **How useful is panoramic vision?** Probably, you will think in terms of being a hunter and being hunted, which remains relevant also for humans. Keep in mind that we humans can see left and right, giving a 180-degree panorama; but what exactly do you see to the left and right? I see only vague shapes and colors there, but I do notice movement there. That is a clue for the even wider panoramic view of animals: *Why is detecting motion useful?* And wouldn't it be useful for humans to also see backward? Why can't we look backward? These are all interesting questions!

Of interest as well is the animals' **sharpness of vision (visual acuity)**. As you know, birds of prey have high acuity; eagles can have two to eight times the human sharpness (with eyes almost as large as human eyes fitting within their much smaller heads); some eagles' eyes also have two foveae that provide very sharp vision in different directions, one straight forward and another about 45 degrees to either side. Pigeons and sharks also have sharper vision than humans.

By contrast, rats and snails have very blurry vision. And it appears that most dogs are near-sighted (and so might enjoy corrective glasses).

Now consider the question: *How useful is sharp vision?* That is easy to answer, but then why do some animals have much sharper vision than others? In particular, why don't land animals have as sharp a vision as birds of prey and sharks?

A number of animals of course have excellent **night vision**. For example, cats may see six times dimmer scenes than we can, while some geckos apparently can see in 350 times dimmer light than humans. For daytime vision, these animals must contract their pupils much more than we do.

How useful is night vision? Again, this question is easily answered, but it also raises many related intriguing issues: Why is there so much diversity in night vision capability among animals? What if all animals had the same degree of night vision? Why don't humans have better night vision and how would such vision change our lives? You may know that modern soldiers have electronic night vision goggles (some can amplify light 50,000-fold, far better than geckos): the advantages in warfare against less well-equipped soldiers are evident; night vision then gives overwhelming and decisive control.

5.1.5 *Facial recognition, neurons and neural networks*

We shall here address briefly a fascinating and complex topic: *How does our brain interpret images seen by our cones and rods?* In Section 5.1.2, we already mentioned the challenge of determining whether a line is straight or not (see Figure 5-4) and the task of identifying a letter (such as the letter "m" in Figure 5-2 or 5-3). Now consider **facial recognition**, an amazing feat of image interpretation: *How does your brain identify people's faces?*

The image of a face on your eye's retina must be distinguished from thousands of other faces that you have seen in person or in photographs, or even in cartoons. The brain must analyze images in a sophisticated way to extract the identity of the observed face despite all the variable lighting conditions, angles of view, hidden parts, glasses, hairstyles, facial expressions, age, pose, make up, *etc*. Recent research has unlocked the mystery of facial recognition by the brain: this is a remarkable achievement, which also and surprisingly has interesting parallels with the 3-dimensional red/green/blue representation of colors that we described in Chapter 4. I will next sketch this story, which involves the **neurons** that form our **neural network** (namely our nervous system), in particular by linking the retina to the brain.

In the 1960s, it was proposed that an individual neuron could respond to the sight of a particular person, such as your grandmother: this type of neuron was thus called **grandmother cell**. In 2005, experiments on patients seemed to confirm the grandmother-cell theory. This finding gave rise to the name "Jennifer Aniston cell", since a single neuron in one patient responded to several pictures of the actress while rarely responding to other images. Another neuron would recognize the face of another celebrity, or a familiar object such as a famous building.

In 2017 a new theory overthrew the grandmother-cell model. The new theory, already verified by experiment, says that a given neuron responds to one of about 50 features of a face (such as details of the shape, dimensions, appearance, and skin folds). The 50 responses from different specialized neurons then together identify the face. With 50 or so distinct features, a huge number of different faces can be distinguished. Your brain must have a store of known faces to compare with and thus identify the person you see. This is very similar to how colors are identified in our brain: certain cones respond to red, others to green and still others to blue; the strengths of those three responses to primary colors define the color that we see, as given in the RGB system with three intensities. So there appears to be no grandmother cell that only responds to the sight of your grandmother, but instead separate neurons that respond to different facial features.

For more information on human facial recognition, I recommend two articles by Diane Martindale[2] and Knvul Sheikh,[3] respectively.

How about machine facial recognition? Nowadays we are familiar with computers that can recognize our faces. Cameras abound in city streets to identify criminals. Border guards often use facial recognition to check people's identities. Your smartphone may use facial recognition to allow you to access it. It should come as no surprise that these systems operate in a way analogous to our brain's facial recognition: they also first measure your face (by evaluating different shapes, dimensions, *etc.*), and then try to find a match with a unique known face.

Such machine facial recognition uses very efficient mathematical and computational methods (called **algorithms**) to identify faces. Some of the most effective methods also use so-called **neural networks** which, as their name indicates, mimic the connections between neurons of our nervous system. These methods are excellent examples of **artificial intelligence**: AI, as it is frequently abbreviated, is a collection of powerful methods that mimic the way our neurons operate. Remarkably, some of these computational methods are more successful than human facial recognition by our brain.

5.2 Film photography

Human vision has one gigantic weakness: we don't remember most of what we saw, the way a picture or a movie can. How much of what you see during a five-minute stroll could you draw faithfully from memory? It is remarkable how little we remember of what we actually see. A related weakness is that we can't show to others what we saw in any detail. Drawings and paintings are just too slow to record and transmit. Photography, first chemical and then digital, brought a solution that is

---

[2] Diane Martindale, *"One Face, One Neuron"*, Scientific American, October 1, 2005, https://www.scientificamerican.com/article/one-face-one-neuron/.

[3] Knvul Sheikh, *"How We Save Face — Researchers Crack the Brain's Facial-Recognition Code"*, Scientific American, June 1, 2017, https://www.scientificamerican.com/article/how-we-save-face-researchers-crack-the-brains-facial-recognition-code/.

now both powerful and cheap. **Film photography** has much in common with our eyes, so we need not repeat the foregoing discussion: for example, **film cameras** have a similar **lens** system as our eyes to focus incoming light onto a light-sensitive chemical film (instead of our retina). We shall here therefore focus on the main differences between our eye's vision and photography.

How did photography start? Photography was first developed in the early 19th century with chemical processes using light-sensitive materials. Initially, photography produced only black-and-white images, but not yet color images, as we will explain below: black-and-white was already very challenging. In the middle of the 19th century, color photography also became practical, although even more challenging.

This approach — often called "film photography" but without implying "movie photography" — still dominated until the late 20th century, even though it was time-consuming: most people sent their exposed film to a processing laboratory for development, especially for color images, and copying pictures was very difficult. Nowadays, film photography is still often used for **movies** to be shown in cinemas and for special high-resolution photography, such as for medical and dental purposes.

However, starting in the 1980s, electronic **digital photography** greatly accelerated and simplified the taking of pictures, although its level of detail was initially poor. The electronic approach has great advantages over film photography, including with color pictures: it allows immediate viewing of pictures, their easy modification, faithful permanent storage and multiple copying, all without loss of quality.

How does chemical film photography work? The first approach to photography was based on chemical changes caused by light. It was found in particular that light of any color would transform tiny silver halide crystals into metallic silver particles (silver halides are salts combining silver with either bromine, chlorine, iodine or fluorine). More light (of any color) produced more silver particles, resulting in a so-called **black-and-white** (or **black/white** or **b/w**) image. A better name would be **gray-scale** image, since the image showed primarily the overall light intensity regardless of color, thus ranging from black to white through all levels of gray in between.

However, the chemical transformation from salt to metal was very slow because it required much light and thus very long exposure times that could be minutes to hours. Therefore, the silver image was usually too weak to see, and certainly not usable for making movies. As a result, it was necessary to "develop" the almost invisible image on film by further chemical reactions to make a stronger image, which was also gray-scale; furthermore, the resulting image must be chemically "fixed" to endure for longer times. Well-equipped professionals and amateurs could perform these reactions in their darkrooms; this still happens with some medical and dental x-ray pictures (x-rays are a form of invisible light): these require developing before you can see your bones or teeth. Most people had to send their light-exposed film to a commercial laboratory for development, which took days to weeks.

For **color film photography**, other chemicals were added to the film to provide sensitivity to different colors. For example, the film could be composed of separate layers with different chemical composition, each layer being sensitive to a different color (such as cyan, magenta and yellow). Again, the immediate image was too weak to see, so further chemical development was needed, layer by layer. Color film processing was much too complicated for most amateurs and required the help of professionals. Color photography only became popular in the 1950s, as the underlying chemical and optical technologies gradually improved both quality and affordability. Weaknesses of film photography include the slow chemical degradation of images over time, the risk of scratching images (think of century-old pictures and movies!), and the difficulty of modifying and copying images.

One notable and popular development around the 1970s was **instant photography**, for example by Polaroid: the film itself contained the chemicals needed to develop and fix the image within minutes after the photograph was taken.

Most color film photography was done either with **negative film** used to subsequently make color prints on photographic paper (thus allowing multiple nearly identical copies), or with **positive film** to make transparent **slides** (but no copies). The small slide images, typically 35×24 millimeters or about 1.4×0.96 inches in size, were normally projected for viewing on screens to do justice to the fine detail available in the pictures; projection screens a few meters in size were often used for group viewing.

For projection of movies in large theatres, where the screens can be ten or more meters in size, sometimes larger film sizes were used (for example, 70 × 48.5 millimeters for IMAX movies), but the detail available with 35 × 24 millimeter film was so good that it was used for most movies. However, after 2000, digital photography largely supplanted film photography for movies, except for the largest cinema screens.

What size of details can be imaged with film photography? The details visible in a film photograph are limited by both the optical lenses of the camera and the size of the film's light-sensitive chemical components, such as its silver salt grains: it appears that about 100 lines per millimeter (or 2540 lines per inch) on the film can be resolved if the optics are very good; this gives 35 × 100 = 3,500 or "3.5K" lines per 35 millimeter image. Such detail in the film translates to details of about 0.05 millimeter (or 0.002 inch) on a standard color print of dimensions 15 × 10 centimeters (6 × 4 inches), or details of about 1 millimeter (0.04 inch) on a screen of size 3.5 meter (12 feet) using a good projector.

This level of detail is comparable to that of high-level digital photography commonly available around the year 2021, which is mainly limited by the size of the individual sensors used to detect the color of separate picture elements or "pixels" (the quality of the optics still plays a role as well): digital cameras around 2021 typically detect about 2,000 ("2K") to 4,000 ("4K") "pixels" horizontally. (We will define pixels in Section 5.3.)

Is photocopying similar to film photography? **Photocopying** is generally meant to rapidly make reasonably good copies of documents like text or images: we thus want it to be quicker and easier than film photography. Photocopying therefore took longer to develop, similar to instant photography. It was only in the 1940s that a practical method of quick photocopying was commercialized. The manufacturer named the method **xerography** (from the Greek words for dry writing) and later adopted Xerox as its company name; nevertheless, Xerox tried hard to prevent the use of the word "xeroxing", common in the USA, in the place of "photocopying"! The technique was initially limited to black-and-white (gray-scale) photocopying.

Xerography consists of two main steps: scanning a document and printing one copy thereof. Briefly, the document to be copied was optically imaged line by line onto a cylinder, charging it electrically where the document was dark. The cylinder was thus covered by an electrical image of the document. A toner (black ink) was then attracted to those electric charges on the cylinder and rolled onto a blank sheet of paper. The toner on the paper was finally pressed and heated for durability. This produced one copy; the whole process had to be repeated for each additional copy.

Color xerography became practical in the 1960s, using similar processes as gray-scale xerography.

Later, digital technology was applied for both scanning and printing in photocopying. Digital scanning and printing are also useful functions for fax machines and computer printers, resulting in "all-in-one printers" that combine scanning, printing, copying and faxing. The main steps are in many ways equivalent to digital photography and printing, as described in Sections 5.3 and 6.1.

Thus, the main differences between photocopying and photography are: (1) xerography uses "dry" writing (without chemical reaction); and (2) the document is scanned line by line (not as one complete snapshot), both in xerography and in digital photocopying.

5.3 Digital photography

Electronic **digital cameras**, just like film cameras, have a lens system to focus incoming light onto light-sensitive detectors: these tiny electronic detectors have the same function as the cones and rods of eyes and the salt crystals of chemical films. However, each electronic detector converts light into an electrical current which is sent to an electronic memory that can store the image element seen by that detector; the many stored image elements together form one picture. An important difference, however, is that digital camera detectors are not randomly arranged as in the eye's retina or in a chemical film: instead, digital detectors are ordered in a regular **grid** (often also called **raster**), usually forming a square or rectangular network covering the entire image.

An example from a digital picture is shown in Figure 5-12, which blows up a small part of Figure 2-1. The entire image is cut up into a regular grid of equal squares: each little square is called a **pixel**, which word is shorthand for "picture element". Every pixel has a single uniform color that is one of the 16,777,216 possible RGB colors. Each pixel comes from a tiny separate detector, which decomposes the incoming light into its RGB components and measures how much red *versus* green *versus* blue that particular pixel contains. This color content is digitized (for example, to specify one of 256 levels for each of the red, green and blue components) and then placed in a digital file for long-term **storing** of that image in a computer. Later, the picture can be manipulated (for example, colors may be changed), copied (without change) and sent to another person's computer (also without change).

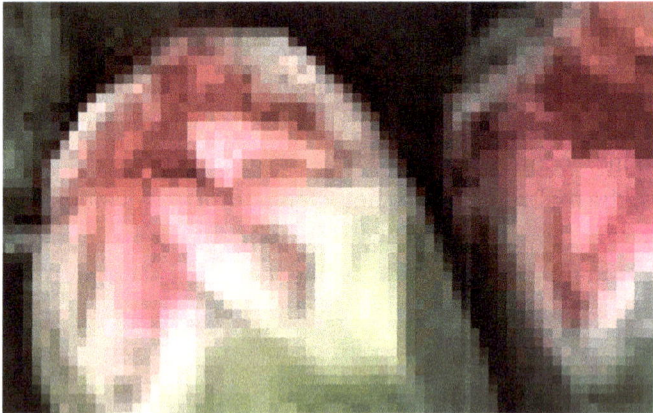

Figure 5-12: Blow-up of my flowers picture, from Figure 2-1, showing individual pixels (about 70 × 40 square pixels are included here, a tiny part of the 2440 × 1264 pixels in Figure 2-1).

What size should the pixels be? **Pixilation**, as the digital cutting up of pictures is called, loses details compared to the original scene, but also reduces the amount of information that needs to be stored and processed. So a compromise is needed: pixels should be as large as possible without significantly degrading the image. As we saw earlier, our visual acuity normally allows us to distinguish dots or lines that are separated by at least 0.1 millimeter (~0.004 inch) at a reading distance of 30 centimeters: that should also be the maximum pixel size in text or images that we want to see at arm's length. Let's consider pictures of digitized text in more detail.

What pixel size is needed to avoid that text become unreadable? Consider the following text, which mixes character sizes, styles and colors:

Times Font 12

Calibri Font 10

Times Font 8

Calibri Font 6

Times Font 4

Calibri Font 2

I printed this text on a standard inkjet printer and photographed it with a smartphone camera. Figure 5-13 shows a close-up of the first line of this text ("Times Font 12"), taken with a simple add-on macro-lens. To set the scale: the height of the letter "T" in that first line is about 0.28 centimeter (~0.11 inch) on the printed paper. The pixels in Figure 5-13 are so small that they cannot be seen at this scale; instead we see the imperfections of the printing process. The letter "T" in this image is 522 pixels high, so each pixel is minuscule: we don't need that much detail to read and it would also create huge digital files, a waste of storage space.

Figure 5-13: This image comes from a highly detailed digital photograph made with a cheap macro-lens attached to a smartphone. The pixel size, about 0.00055 centimeter (~0.0002 inch) on the original paper, is too small to see even at this magnification: the height of the letter "T" on the original page is about 0.28 centimeter (~0.11 inch), and is 522 pixels in this digital image.

We now consider larger pixels: I have changed the pixel size (with graphical software) to produce the top image in Figure 5-14. Now the pixels are visible as small squares of variable intensity, especially along the edges of the letters (to clearly see the pixel squares, look at the blowup of the first two lines in the bottom part of that figure). The same letter "T" is now about 28 pixels high and still clearly readable, including from a distance where you don't see the individual pixels. Also shown in Figure 5-14 are lines of text in smaller character sizes (called smaller font sizes) and partly in the simpler Calibri font. The pixel size is here about 0.01 centimeter ~0.004 inch on the original paper.

Figure 5-14: Top image: As Figure 5-13, but with larger pixels of about 0.01 centimeter ~0.004 inch in height on the original paper: the height of the letter "T", still in font size 12, is now about 28 pixels. The additional text has font sizes 10, 8, 6, 4 and 2, as indicated. Font Calibri alternates with font Times New Roman. The colors are R, G, B, then C, M, Y, in that order. The bottom image blows up part of the top image to better display the pixel squares for font sizes 12 and 10.

YOU MAY WONDER: ***What exactly are font and font size?*** First, **font** is a style of character used in printing (a font is also often called a typeface), such as the common styles Times New Roman and Calibri (word processors offer tens of different fonts to choose from); in addition, you can choose **bold** or *italic* styles. Second, the size of the printed characters needs to be specified by means of a **font size**: by definition, the font's full height (including the parts extending above and below the character "x") is given in points, where one point is 1/72 inch ~0.035 centimeter. So, the font size 12 indicates a full character height of 12/72 = 1/6 inch ~ 0.42 centimeter. The capital letter "T" does not extend below the letter "x" and has a height of about 2/3 of 0.42 centimeter, namely 0.28 centimeter ~0.11 inch. Other font sizes have letter heights and widths that are simply proportional to the given font size: so font size 6 has characters that are half as large as font 12.

You can see in Figure 5-14 how readability decreases for smaller font sizes and larger pixel sizes: if you watch from farther away to avoid seeing the pixels, you can then no longer properly identify the letters. On the other hand, the gain in storage space is enormous: from 255 to 28 pixels means a reduction in storage space by a factor $(522/28)^2$ ~ 348 (the square comes from the two dimensions of the image); in other words, you can store 348 times more images in the same space using the larger pixels, and still read smaller characters, perhaps down to font size 8 if you have good eyes.

It is interesting to compare these pixel sizes with our visual acuity: as mentioned in Section 5.1.2, with normal vision and a reading distance of 30 centimeters, you can distinguish dots separated by about 0.01 centimeter (0.1 millimeter ~0.004 inch). This means that we can just distinguish the pixels in the text that is pictured in Figure 5-14. **Thus, our visual acuity matches well the recommendation of font size 12 for good readability.**

Can we confirm the above conclusions with idealized characters? The discussion above looked at actual printed characters through an actual camera. So, we may wonder whether we can, for our discussion, avoid the limitations of real printing and real photography by simulating the effect of pixilation with "perfect" computer-generated characters. We will do so by taking screenshots instead of printing and photographing:

I will create "perfect" letters in a word processor and capture their images directly within the computer, so that we can concentrate our attention on the pixilation alone.

As we saw above, when the pixels become larger, the letters and other characters become more jagged and more difficult to identify. This is illustrated again in Figure 5-15, where the letter pair *"AG"* is shown

1024x512 pixels

8x4 pixels

512x256 pixels

16x8 pixels

256x128 pixels

32x16 pixels

128x64 pixels

64x32 pixels

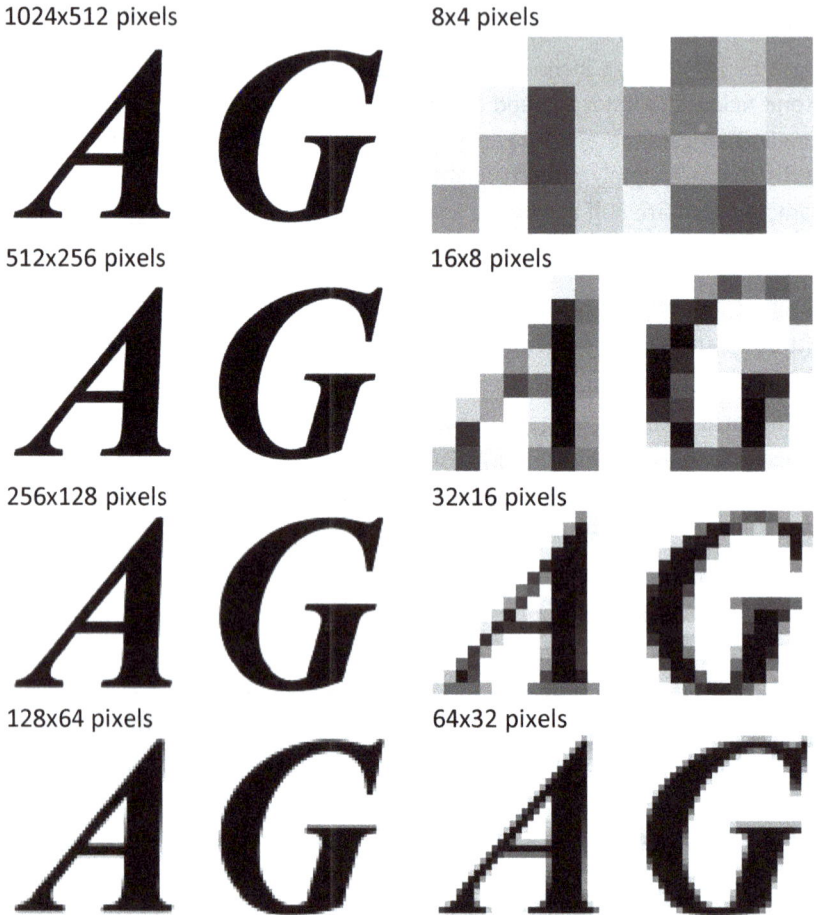

Figure 5-15: The letter pairs *"AG"* in font Times New Roman are shown here eight times with equal heights, but increasing pixel sizes: the pixel size doubles each time going down the left column and then up the right column (the notation 1024 × 512, for example, indicates a width of 1024 pixels for the pair of characters taken together, while their height is 512 pixels). The pixels along the edges of letters are gray: the gray level of a given pixel is the average over the original smaller pixels inside the larger given pixel.

eight times with different pixel sizes. Going down the left column and up the right column, the pixels are doubled in size each time. From a letter height of 512 pixels (top left, very close to the 522 pixels in Figure 5-13) to 128 pixels (third down in the left column), the letters look perfectly sharp to the naked eye. But a height of 64 pixels (bottom left) shows jagged edges around the letters; this jaggedness is strongly accentuated for heights of 32, 16, and 8 pixels (right column); letters with a height of 4 pixels are practically impossible to recognize.

In Figure 5-16, we show again the same pair of letters "*AG*" eight times, but now repeatedly halved in size so that the pixel size remains the same. Again, the largest letter pair is 512 pixels high, while the smallest pair is 4 pixels high (it is barely visible as a blurry dot in the center of the spiral of letters). The individual pixels are small enough to be invisible to the naked eye (you may magnify the figure, but the

Figure 5-16: Rearrangement of the eight letter pairs "*AG*" of Figure 5-15. The top letter pair "*AG*" is repeated seven times in an inward spiral, halving its size each time, so that the seven smaller letter pairs are 1/2, 1/4, 1/8, 1/16, 1/32, 1/64, and 1/128 the size of the top pair. All letter pairs use the same pixel size, which is 1/512 of the largest letter height, and 1/4 of the smallest letter height.

computer display will probably blur the pixels so they remain invisible: can you read the smallest 4-pixel-high letters? You may be surprised how well the brain can extract letters from a blur!).

We can thus confirm that our earlier findings about text readability are valid: both pixel size and character size are of primary importance.

In Figure 5-15, we see again that each pixel has a uniform color. Although these pixels represent black-on-white text, they are not necessarily black or white: the pixels on the letters' edges are gray. This shows that the pixels take the average of the colors that they cover, so that a pixel on a letter's edge will have an average color between black and white, namely some level of gray: that pixel's gray will be darker if the pixel covers more of the black letter, and it will be brighter if it covers more of the white background. The same is true in Figure 5-12: the color of a pixel is the average color of the original scene everywhere inside that pixel. (This averaging also occurs with the cones and rods in the eye and the salt crystals in chemical film.)

How is color detected in digital photography? In every pixel, a camera must determine how much of each R, G and B color component arrives from the pictured scene. The most common approach is to use CMOS electronic light sensors (CMOS stands for "complementary metal-

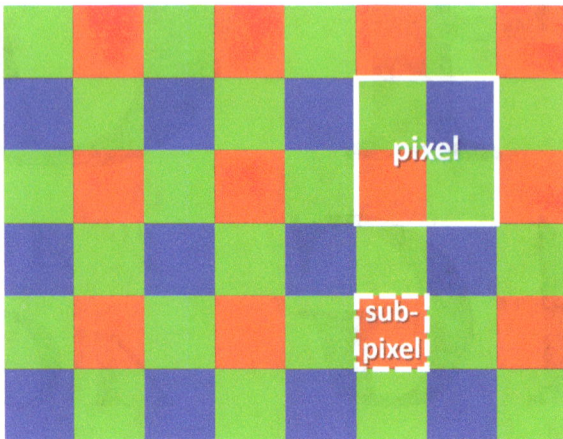

Figure 5-17: Bayer pattern of color filters (subpixels) used in RGB color digital photography. Twice as many green filters are present compared to red and blue filters: this is because the human eye is more sensitive to green than to red and especially blue. Each square group of four filters (two green, one red and one blue) forms a pixel.

oxide-semiconductor" technology). Each sensor is covered with a color filter; the color filters are typically arranged in groups of four sensors per pixel, ordered in a **Bayer pattern**, as illustrated in Figure 5-17. Each color filter allows mainly one color (either red, green or blue) to pass to the underlying CMOS sensor. The different color intensities from the four sensors in one pixel are measured as four different electrical current intensities, which then are typically translated into three R, G and B intensities from 0 to 255. (One filter and one sensor together form one subpixel.) Typical pixel sizes range from about 1 to about 8 micrometers (from ~0.04 to ~0.32 mil); for comparison, these pixel sizes are about 2 to 16 times larger than the cones in the human eye's fovea.

> CAUTION: The word "pixel" is sometimes used to mean an individual sensor rather than a group of four sensors described above. In this book we always use pixel to mean the smallest part of an image which can be visually distinguished: thus, here a pixel is one group of four sensors, each of which contributes one "subpixel".

An alternative approach, the Foveon X3 sensor, operates more like chemical color film: sensors are arranged in different layers for different colors; the layers serve to both filter colors and measure light intensities. Thus, each pixel contains three sensors at different depths, such that, for example, the top sensor gives blue light intensity, the middle sensor gives green light intensity, and the deepest sensor gives red light intensity. (More precisely, because of the layer-by-layer filtering, the top sensor measures all light, the middle sensor measures green and red light together, while the deepest sensor measures only red light: from these measurements, the separate blue, green and red components can be calculated.) The size of each pixel is typically 7 micrometers (~0.28 mil) on a side; this pixel size is about 10 times larger than the cones in the human eye's fovea.

Do all cameras produce the same colors in pictures every time? The short answer is no! The color measured by a camera depends on several factors: imperfect lenses shift colors differently in the image, in what is called chromatic aberration, and absorb colors differently; color filters have varying chemical compositions that give different weights to different colors; sensors also have varying chemical compositions that result in different sensitivities to different colors; different algorithms (calculation formulas) are used to process the measured pixel color, including for example **white balance** (mentioned in Section 2.1) and many other aspects of a picture. Therefore: Different brands and models of cameras will give different picture colors; also, different lighting conditions will result in different picture colors.

IF YOU HAVE DEALT WITH IMAGE FILES BEFORE, THE FOLLOWING TOPICS MAY BE OF INTEREST TO YOU: *How are digital images stored in computer files?* You are probably aware of some of the file types commonly used for digital pictures: examples of digital file formats for images include **JPEG** (which gives the suffix .jpeg or .jpg for file names, as in mypicture.jpg), **GIF** (as in mypicture.gif) and **Bitmap** (as in mypicture.bmp). The Bitmap format simply lists the color of each pixel as RGB numbers, often producing large files because there are millions of pixels in typical images, as we will see below. The **RAW** format is similar to Bitmap, and indicates that the image data in the RAW file were not processed by the camera before storage (the data are thus literally "raw"); however, the RAW format is not unique as it varies between camera manufacturers.

JPEG and GIF use **data compression** to reduce the **file size**. JPEG exploits the similarity of smoothly varying colors across pixels and uses that relationship to greatly reduce the amount of stored information (more precisely, for the mathematically inclined reader: JPEG fits smooth cosine functions to the image). GIF uses the fact that, in many images, large areas have exactly the same color. This is especially common in pictures of text, diagrams and drawings. For example, on a white page with black text, the large majority of pixels are white, while only some are black (within the letters) or gray (on the edges of letters): that information can be stored efficiently to avoid listing all white and black pixels individually (for example, writing "1000 × 255255255" takes much less space than writing "255255255" one thousand times to indicate one thousand consecutive white pixels). GIF is "lossless", meaning that no information is lost in the compression and the original image can be recreated exactly. However, JPEG is "lossy", meaning that some details are lost, and the original image cannot be recreated in all details (especially sharp lines and abrupt boundaries).

What is the size of typical digital photographs? The size of a digital picture can be counted as the number of pixels into which the camera has cut it up. For example, a camera may have 4032 detectors horizontally and 3024 vertically: this implies 4032 × 3024 = 12,192,768 pixels altogether, which is close to 12.2 million and therefore may be called 12.2 **megapixels** or 12.2 **MP**. The higher is the number of pixels, the more detail the image will contain. Typical cameras around 2021 produce image sizes between 1 and 20 MP; this size increases steadily in new cameras, but has already reached levels of detail that are more than sufficient for good pictures in everyday life. As an interesting comparison, the complete retina in the eye has about 6 million cones and a smaller number of rods, which is similar to 6 megapixels. However, you will need more pixels (which means more detail) if you want to magnify a digital picture for displaying on a large-screen television or printing on a large advertisement board, for example, or if you want to **crop** (meaning cut out) and then magnify a small part of a picture, such as a face.

A related question is the **file size** of a picture: the file size determines how many pictures you can store in your computer, how many pictures you can attach to your electronic messages and how fast your messages will travel over the internet. File size refers to computer storage, which is usually measured in **bytes** and **megabytes** (a megabyte is a million bytes or 1 **MB**). A byte contains one number, for example the intensity level of an RGB color, such as the number 255 for the maximum intensity of green. Thus, each pixel requires 3 bytes: one byte for each RGB color component. Assuming $4032 \times 3024 = 12{,}192{,}768$ pixels, we need $3 \times 12{,}192{,}768 = 36{,}578{,}304$ bytes or about 36.6 MB of storage for one photograph. This value of 36.6 MB is the storage space needed using the Bitmap file format (.bmp). It is also the size of the email attachment in which you may try to send that picture: more than likely that email will be rejected as being far too large!

(In practice, the file size in your computer can be a little bit different for two reasons: first, the file will contain some additional information, so-called metadata such as date, time, camera type, lens settings and even location where the picture was taken; second, the term megabyte sometimes means 1,048,576 bytes rather than 1,000,000 bytes, because 1,048,576 is the twentieth power of 2, which is convenient for computer design: therefore, there can be a difference of about 5% between quoted sizes in megabytes.)

The compression techniques like those used in the JPEG and GIF format are very useful to reduce file sizes and send pictures by email: for example, the original flower picture of Figure 2-1 is reduced by a factor of about 6 to about 6 MB in both JPEG and GIF formats; a less complex picture, like Figure 2-24 is reduced even more, to about 1.8 MB. Indeed, the compression depends very much on image complexity. These sizes can be reduced even further, by almost imperceptibly sacrificing a bit of accuracy in colors, to 0.2 MB for Figure 2-24, for example. A typical photographed text page can be compressed to about

(Continued)

(*Continued*)

0.2 MB from 36.6 MB in the GIF format; by using fewer (larger) pixels, it can be further reduced to less than 0.1 MB (0.1 MB is 100 kilobytes or 100 KB).

When reducing the picture file size by shrinking the number of pixels and thus enlarging the pixel size, we sacrifice detail; depending on the picture and how it will be viewed, this may be acceptable. For example, we can halve the number of pixels in both the horizontal and vertical directions (this is called **resizing** or **resampling**): this reduces the total number of pixels fourfold (from 12,192,768 to 3,048,192 pixels) and also reduces the file size fourfold (to 9.15 MB for Bitmap or about 1.5 MB for JPEG and GIF); further reductions are again possible by sacrificing accuracy. Converting images between formats, levels of accuracy and number of pixels, as described here, can be done with commonly available graphics software, some being free.

5.4 What have we learned in this Chapter?

We read text and scrutinize images primarily with one small part in the center of the field of view of our eyes. Our visual acuity, and thus our sharpness of view, is limited mainly by the size of our eyes' cones and by our eye's focusing ability. While we normally detect three colors (red, green and blue), some animals detect one or two colors, and others four or even more colors.

Photography evolved from a slow chemical process that produced only black-white (gray-scale) images to a fast and cheap digital technology. Digital images are composed of detector pixels that define the detail and colors recorded. Different brands and models of cameras give different picture colors; also, different lighting conditions result in different picture colors.

6

Creating Color and Displaying Color

After recording color images, we wish to display them for viewing on paper or canvas, or on electronic monitors. We may also like to create a color image or text by drawing it from scratch. This requires both **creating color** *and* **displaying color**.

In this Chapter, we first discuss the case of making an image on paper or canvas: this relies on light reflected from ink or paint. Apart from painting, the more common method is currently printing with an inkjet printer, which lays down very small CMY-colored dots.

Next, we discuss showing images with electronic displays, such as computer monitors (including smartphones) and television screens. Here also, small dots are produced (but with RGB colors instead of CMY colors): these dots are called pixels. We will show how, with commonly available software and hardware, you can easily produce nearly 17 million colors.

We will also discuss the variety of sources of light that are available, from sunlight and household lamps to plasma TVs and lightning.

— ·)⟩⟩ ⟨⟨· —

6.1   Creating and displaying ink and paint color

As mentioned in Section 4.1, we can use the subtractive process of the **CMY color system** to create color with **ink** or **paint**, for **printing** and **painting**. This approach starts with the white background of paper or canvas, which is due to incoming white light. After all, without incoming light, you can't read printed text or enjoy a painting. (Using non-white paper or canvas, or non-white incoming light, gives further interesting possibilities, which I leave to your imagination or your analysis of reflected colors, as discussed in Section 4.2!)

   How can we remove colors from white to achieve a desired color? Can you think of ways to remove certain colors from light using ink or paint? How about with color filters? Indeed, to remove colors from white, the ink or paint can act as a color filter. To show how this can be done, in Figure 6-1 we imagine white light falling on a white surface, like paper (a cloth surface would work essentially the same way, without being as flat as paper). Our approach will be very similar to our discussion of colors reflected from surfaces in Section 4.2, but here we will consider thin layers of ink or paint, and we will focus more on CMY colors, while it remains convenient to think in terms of the RGB components of light.

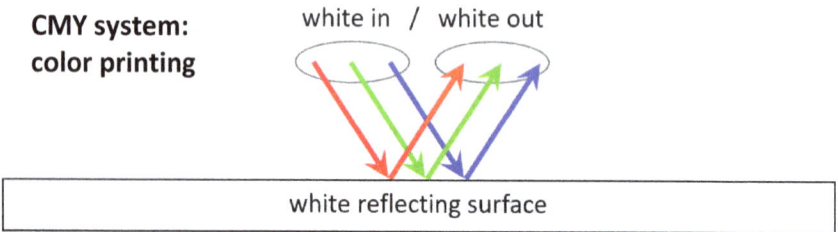

Figure 6-1: Reflection of white light by a white surface: the red, green and blue components of white light are all reflected.

   On the unprinted white paper, the three primary colors red, green and blue reflect equally well, and together therefore produce white reflected light, as already shown at top left in Figure 4-5. This white surface corresponds to the white background around the three circles in the CMY diagram of Figure 4-1.

   Let's add a layer of ink onto the paper, such as the cyan layer shown at left in Figure 6-2: if the layer has the proper thickness, this ink absorbs red light but allows green and blue to go through. Now the green and blue will reflect from the underlying paper, pass again through the ink layer and together produce cyan reflected light: this red-absorbing ink layer will thus look cyan in color. Similarly, another type of ink layer can absorb green and let through red and blue, producing a magenta reflection, while a layer that absorbs blue will let pass red and green, giving a yellow reflection. These single layers of ink correspond to the non-overlapping parts of the three circles in the CMY diagram of Figure 4-1, where we see cyan, magenta and yellow colors.

By adding a second ink layer (see Figure 6-3), we can remove more colors. If one layer removes green and another removes blue, only red is left to be reflected. Other combinations produce green or blue reflected light, as illustrated. These double layers of ink produce the red, green and blue parts of the three circles in the CMY diagram of Figure 4-1, where the circles overlap pairwise as double layers.

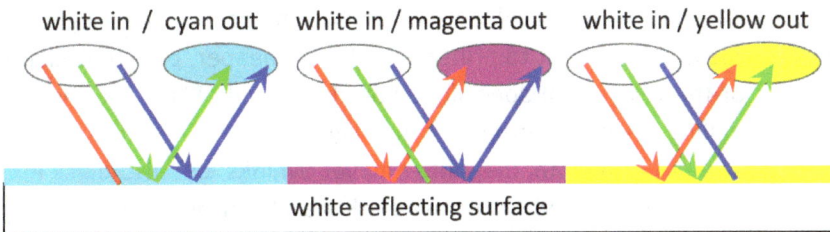

Figure 6-2: Reflection of white light by a white surface covered by a layer of cyan (left), magenta (center) or yellow (right) ink: in each case, one of the red, green or blue components of white light is absorbed, while the other two are reflected.

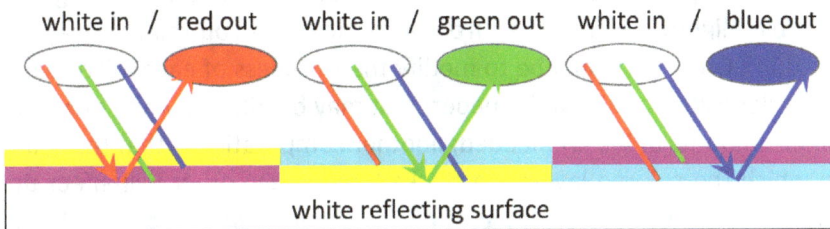

Figure 6-3: Reflection of white light by a white surface covered by two layers of cyan, magenta or yellow ink: in each case, two of the red, green or blue components of white light are absorbed, while the third component is reflected.

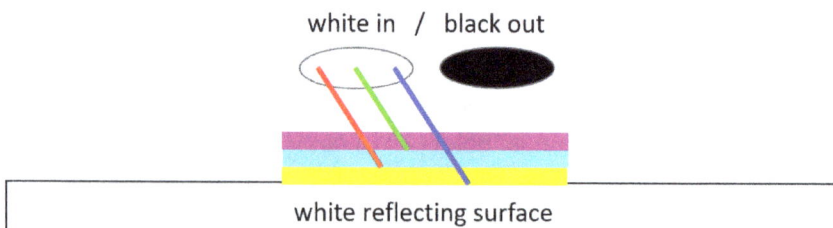

Figure 6-4: Reflection of white light by a white surface covered by three layers of cyan, magenta and yellow ink: each layer removes one of the red, green or blue components of white light, such that no light is reflected, thus producing black.

Most printing produces black text, not color. ***How can we print in black? Can you think of a way to print black with the CMY system?*** Here is one way: black can be achieved by adding a third ink layer; the three ink layers together remove all colors, leaving no light, namely black. This is shown in Figure 6-4, and also in the center of Figure 4-1: there, three ink layers remove the three primary colors, leaving no light.

This approach to making black, however, is expensive and contradictory: it requires a lot of color ink to produce no color (and no light)! It is of course more efficient to print one layer of cheaper, purely black ink instead of three layers of specially tailored color ink. Black ink absorbs all light falling on it, regardless of the light's color. Therefore, inkjet printers use not only cyan, magenta and yellow inks, but also black ink: this color system is called CMYK, where K stands for black (B is already used for blue, so K is chosen instead). Some printers use different black inks for pictures as opposed to text, so they need five ink cartridges.

How can we make other colors than the basic colors with CMY inks or with paints? To make other colors than discussed above in the CMY color system, we need to be able to vary the amount of red, green and blue light being reflected from a surface. Can you think of ways to do that? One way would be to modify the thickness of each ink or paint layer deposited on the white paper; that may be difficult when only very small amounts of ink are needed; another complication is that the order in which the colored layers are stacked upon each other will affect the final perceived color. The easier and more common approach is to lay down ink as tiny colored dots with varying sizes or density. To produce a color containing more cyan, for example, larger cyan dots are printed on a grid. This approach is illustrated in Figure 6-5. In the upper three

Figure 6-5: Blow-ups of parts of images with CMY colors printed as dots on white paper, in regular grids (top three images) or irregular arrays (lower images). The bottom row shows red, orange, yellow, green, cyan and blue parts of the simulated solar spectrum of Figure 2-11. The last-but-one row shows black, mid gray and light gray from the black-to-white color bar of Figure 2-14, including a strip of bare white paper at far left. The structure of the paper itself is also visible in these images as streaks or "folds". (The large image in the center comes from the same flower picture as Figure 2-1, after printing on paper.)

images we see dots arranged in square grids, each grid being oriented according to its color (cyan, magenta, yellow or black) so as to minimize the overlap of dots of different colors; such grids are often used to print images in mass publications.

The lower three rows of images are from printouts of an **inkjet printer** which aims jets of tiny ink droplets onto the paper in a disordered array (note that the gray colors in the one-but-last row actually include colored dots in addition to black dots). Of course, the dots are kept very small and close together, so that our eyes cannot distinguish them: we just see the smooth combined average color; my photograph of an image with an eye in Figure 6-5 is out of focus around its edges, showing this averaging and smoothing effect.

You can explore such printing techniques yourself in newspapers, glossy magazines, books, brochures, *etc.*, using a good magnifying glass. You can also use your own color printer and examine its output.

With paint one normally does not use dots, except in some forms of artistic painting such as **pointillism**, where the same effect is obtained with irregularly positioned colored dots instead of continuous smooth colors. Direct mixing of the liquid paints achieves the same result, as we discussed in Section 2.2 after the question *"How does mixing paint work?"*

6.2 Creating and displaying digital color

Section 6.1 explained how we can produce color with ink and paint by using the subtractive process of the CMY color system, starting with a white surface. In the case of **electronic displays** (computer screens, television sets, smartphones, *etc.*), the additive process of the RGB color system is more appropriate, as discussed in Section 4.1. This additive process means that we start with a black display (it emits no light) and then turn on emitters within the display to produce light of various colors. This is of course the reason that you can use an electronic display in the dark, unlike a printed or painted surface: the display produces its own light, while ink or paint can only reflect external light if it exists.

The contrast between these approaches leads to a surprising but very useful recommendation: To minimize the recharging of smartphones, tablets and pads, we should display text as white on black (or blue on on black), not as **black on white** (or blue on white)! This is the opposite of printing with ink, which consumes less ink when we print black text on white rather than white text on black. Indeed, electronically displaying text as black on white consumes much electricity to produce all that white background; it is therefore much more economical to light up the tiny area of white letters (as allowed by some smartphones). *Can we get used to reading* white-on-black text *? Or do we prefer black-on-white text over white-on-black text, and why?* You decide.

6.2.1 *Emitter pixels*

How does an electronic display produce the desired colors? Maybe you have looked very closely at a computer monitor or television screen and seen something like the images of Figure 6-6: these are close-up photographs of a part of the flower picture in Figure 2-1, as displayed on a computer monitor. The left image clearly shows colored squares: these are the individual light emitters, forming one picture element or **pixel** for each emitter; each emitter produces one particular color, in this case some shade of red.

The right image in Figure 6-6 is a separate, more detailed photograph of the blue boxed section within the left image (it is not a simple magnification of the left image): now we see red, green and blue bars of various intensities within the colored squares. Each square pixel (that is, each picture element outlined in yellow in the image) consists of three such bars. Each bar is a sub-emitter that produces either red, green or blue light. Such a sub-emitter has a specific fixed chemical composition that is responsible for its color being either red, green or blue; changing the voltage applied to a sub-emitter controls its brightness. As we will discuss in more detail in Section 6.2.2, the overall color of a pixel is determined by the different intensities of light emitted from the three sub-emitters in that pixel, following the RGB system that we introduced earlier.

Are the squares in Figure 6-6 simply the pixels of the original digital photograph shown in Figure 2-1? To answer that question, let's compare the two images side by side: in Figure 6-7, the left image is the same as in Figure 6-6, while the right image is extracted from the original photograph shown in Figure 2-1. If you look closely, you will see that the

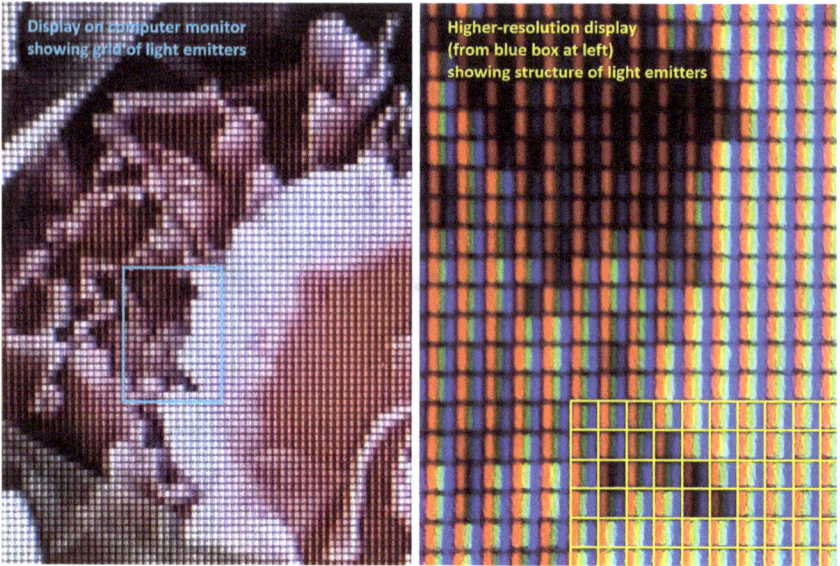

Figure 6-6: Close-up photographs of a part of Figure 2-1 as displayed on a computer screen: the right image details the blue box of the left image. The squares in the left image correspond to light emitters, while the right image shows the internal red/ green/blue structure of individual emitters outlined in yellow. (To see such detail on a computer monitor it helps to use a magnifying glass, while many cameras and some smartphones can make such close-up photographs, especially with a macro setting or lens, as I did with a smartphone.)

right image in Figure 6-7 is also made up of squares (these are the pixels which we mentioned in Section 5.3, which correspond to the detectors in the camera). But these squares are smaller than the squares in the left image (which are the emitters in the display): in this example, the emitters (at left, highlighted in blue) are about twice the size of the pixels of the original photograph (at right, highlighted in yellow); compare them side-by-side at bottom. This is not surprising when we realize that the original photograph was taken with 4032×3024 detector pixels, while the same picture was displayed on a computer monitor that has only 1600×920 emitter pixels.

We conclude: **A computer display does not simply show the original pixels of the digital photograph.** The reason is simple: the grid of light detectors in a camera is usually <u>different</u> from the grid of emitters in a display. In the case of Figure 6-7, the camera had 4032×3024

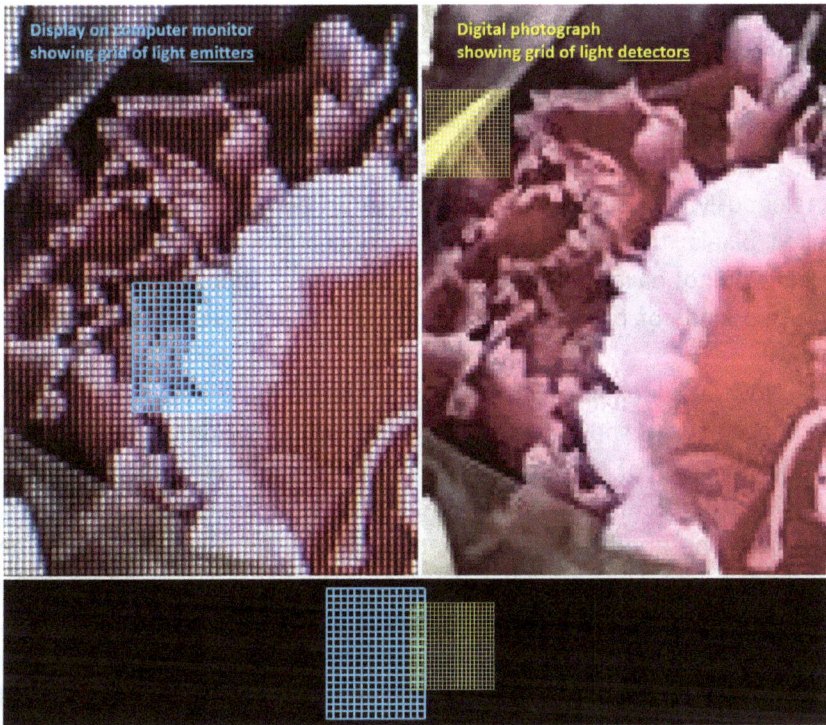

Figure 6-7: The left image is the same as the left image in Figure 6-6, while the right image is extracted from the original digital photograph shown in Figure 2-1: here its pixels are visible. The yellow grid shows the grid of original pixels at right, while the blue grid shows the grid of squares at left. Both grids are copied to the bottom to compare their sizes: they are clearly not simply related to each other!

detectors, while the monitor had 1600×920 emitters. In general, different cameras have a variety of numbers of detectors, while different monitors also have a variety of numbers of emitters: it is therefore very rare for the digital photograph to fit precisely square-for-square on the display's grid. (Besides, some cameras and displays don't use squares but other shapes.) This mismatch, if not corrected, leads to very distracting artificial wavy bands called moiré patterns, which are described in Chapter 9.

As a result, **the grid of pixels of the digital photograph must be converted to the emitter grid of the display**: this is done automatically by the graphics card in a computer or television set, using a technique called interpolation (which means calculating the expected color at a point in between other points where the colors are known, for example by averaging over the four nearest squares, weighted by how close those squares are).

A note about terminology: We have mentioned the pixels of a digital photograph, which correspond to the detectors in a camera. Analogously, the squares or emitters in an electronic display are also called **pixels**: this may be confusing, since these two kinds of pixels usually do not have the same functions or sizes. I therefore differentiate the two kinds of pixels by calling them **detector pixels** *versus* **emitter pixels**.

Now let's look more closely at the images in Figure 6-6. The close-up image at right shows the internal structure of the emitters contained in the blue box of the left image. The yellow overlaid squares outline individual emitters. Each emitter is seen to consist of three vertical bars: one red, one green and one blue. These are three sub-emitters: one emits red light of some intensity, the next emits green light of some other intensity, and the third emits blue light of yet another intensity. This may remind you of the RGB system. Indeed: The three sub-emitters correspond to the RGB colors and together they can produce any of the roughly 17 million RGB colors.

For example, in the blue box of Figure 6-6, we have a range of whites and reds. The right edge of the blue box is almost white: the corresponding emitter pixels in the right image therefore emit relatively intense red, green and blue to produce an almost white color. Near the bottom of the blue box are a few almost black points: the corresponding emitter pixels indeed emit close to zero red, green and blue, as we can see at right in the yellow-highlighted section. Other areas toward the left are predominantly red: so the red sub-emitters are relatively brighter there; notice how weakly the green and blue emitters light up near the top of the right image.

What do the pixels for simple basic colors look like on a display? Figure 6-8 shows three photographs of a display with uniform patches of the six basic colors (red, green, blue, cyan, magenta, yellow), as well as white. We see how the three sub-emitters provide the necessary color components to make the colors cyan (green + blue), magenta (red + blue) and yellow (red + green), as well as white (red + green + blue). This additive mixing reminds us of the RGB color circles of Figure 2-13, where cyan, magenta, yellow and white were obtained by mixing red, green and blue.

The sharp boundaries between color patches in Figure 6-8 show clearly how the red, green and blue sub-emitters line up from one color patch to the next.

To see the compound colors cyan, magenta, yellow and white properly, you have to watch from a distance or shrink the figure, so that you can't distinguish the pixel structure anymore.

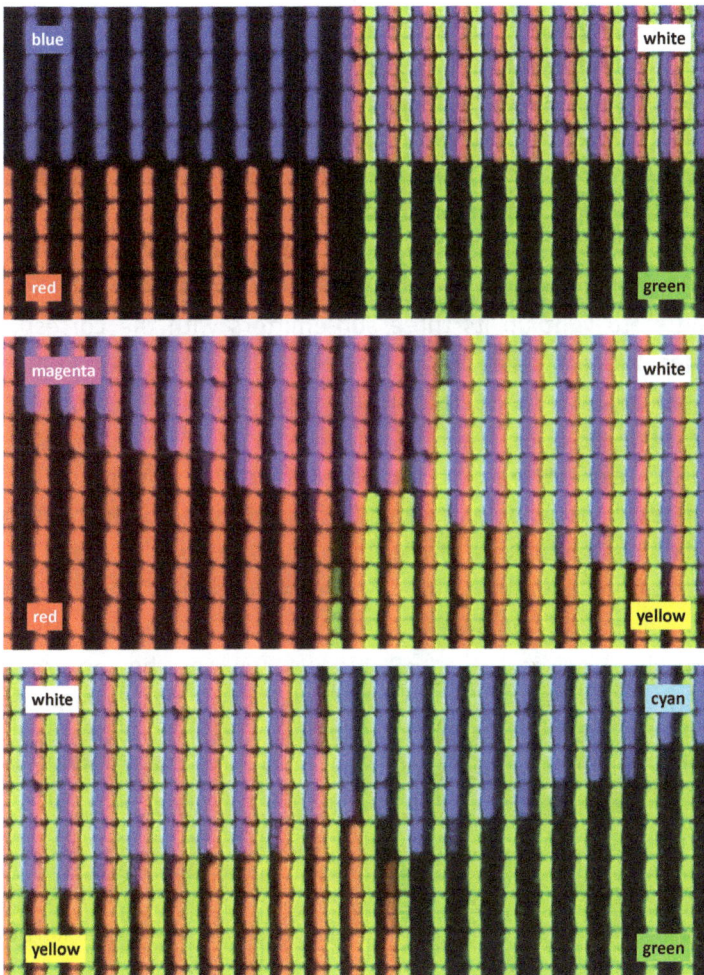

Figure 6-8: Three closeup photographs of a computer screen displaying pure basic colors.

You may wonder why these colors, when seen from a distance, look relatively dark; this is especially obvious when looking at the "white" patches, which appear dark compared to the whiteness of the background page around the figures. We will discuss this aspect later in this Section. We will then also address the relatively high intensity of green and relatively low intensity of blue compared to red.

Does the structure of the emitter pixels correspond to the structure of the detector pixels? We saw in Figure 5-17 the typical structure of a detector pixel: a Bayer pattern of two green sensors, one red sensor and one blue sensor. However, a typical emitter pixel differs (apart from its size) in having three equal vertical bars, as seen in Figure 6-8. So: There is no direct connection between the structure of detector pixels and emitter pixels. The connection is indirect: the four sensors in a detector pixel with Bayer pattern produce signals that lead, *via* some computation, to the red/green/blue composition of that pixel (for example R = 255, G = 255, B = 0 for a pixel detecting a yellow color); this RGB composition is then passed to the graphics card which instructs the emitter to light up its three sub-emitters with the corresponding intensities (the interpolation mentioned earlier in this Section must first be applied because the detector pixels and emitter pixels probably don't match up one for one).

How is text written on a computer screen? My photo in Figure 6-9 gives an example: the letter pair "ay" is displayed with a height of 16 pixels. This reminds us of the letters *"AG"* shown in Figure 5-15; in fact, the letters "ay" here have the same number of vertical pixels as the letter pair *"AG"* labeled *"32 × 16 pixels"* in Figure 5-15. The letters "ay" are also black in Figure 6-9, but we now see the red/green/blue sub-emitters instead of just gray-scale detector pixels. A gray level in an emitter pixel is achieved by producing equal light intensities from its red, green and blue sub-emitters.

We need to address two puzzles about the white color in the RGB system. We have repeatedly claimed that white can be represented as the combination of the same amount of pure red with pure green and pure blue. However, when we look at the "white" background of Figure 6-9, for example, green looks clearly brighter than red and much brighter than blue: why does their brightness appear so different? This contrast was already visible in Figures 2-13 and 6-6.

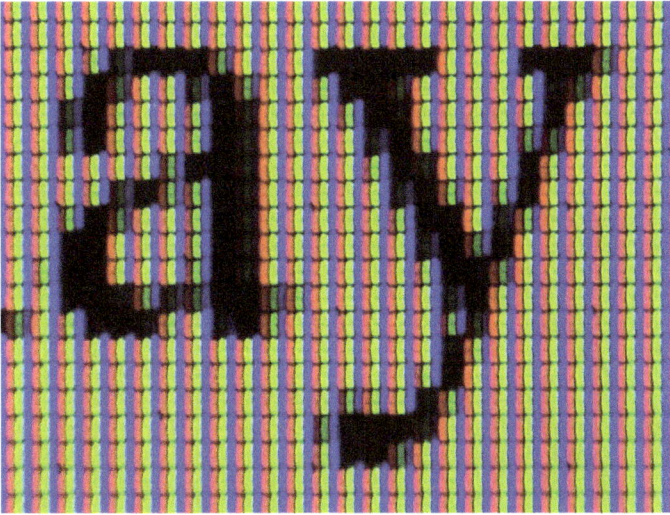

Figure 6-9: Photograph of the black letters "ay" on a white background displayed on a computer monitor.

The cause of these large differences in <u>perceived</u> intensity is the difference in **sensitivity** of our eyes' red *versus* green *versus* blue cones. Indeed: Our green cones are about two times more sensitive than our red cones, and six times more sensitive than our blue cones. So we see red and especially blue as weaker than green. (We can speculate why that would be: perhaps early humans dealt mostly with green vegetation, saw relatively less red meat or yellow fruit and very little blue fruit, while the blue color of the sky was less important to them.)

The second puzzle is the following: Consider, for example, the white colors in Figure 6-9 or Figure 6-8. *Why does the eye see the three sub-emitters (red, green and blue) clearly darker than the white that they produce together?* In brief, this is due mainly to three effects: (1) magnifying an image to see the small sub-pixels spreads the light over a larger area, making it darker; (2) a camera's time exposure is automatically adjusted to provide a balanced color mix, which in this case also means darkening; and (3) a camera also automatically adjusts the so-called **white balance**, which again in this case darkens the picture. Therefore: The colors in a close-up photograph may be significantly different from the colors in a non-close-up photograph. The following note gives a more detailed discussion of these effects.

HERE ARE MORE DETAILS ABOUT THE THREE EFFECTS MENTIONED ABOVE: The first effect is that, when we magnify an image to see sub-emitters, we are spreading the light intensity over a larger area, so making it darker. Therefore, when we magnify a white surface, the white level is reduced to gray, and all other colors are also darkened. The second effect is the time exposure that the camera automatically selects to achieve a reasonable overall intensity in the picture: this effect depends a lot on other parts of the picture that we may not be interested in, because the camera evaluates and adjusts the overall light intensity of the entire scene. The third effect is the camera's automatic white balance: the camera's software is programmed to adjust the colors to achieve a normal "solar" white, for example in order to compensate for artificial illumination with non-white light sources. However, this requires the camera to "guess" what the illuminating color actually is; the result can vary a lot from camera to camera and from scene to scene (as an example, see the plastic bag in Figure 2-2 which looks bluish but in reality was white).

This situation is schematically represented in Figure 6-10: one emitter pixel is shown producing the same maximum amounts of red, green and blue (remember that 255 is the maximum intensity of light in the RGB system, using an intensity scale from 0 to 255); this produces white in our eyes. In our following discussion of how to produce colors, we will use the representation of Figure 6-10 to connect the RGB color

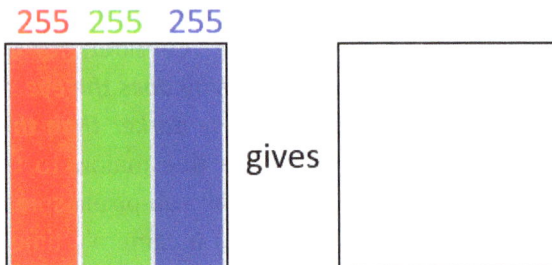

Figure 6-10: The left square represents one emitter pixel consisting of three sub-emitters producing the maximum allowed amount (RGB intensity level 255) of red, green and blue light. The real sub-emitters actually emit about three times more light intensity than is represented here, so as to produce the white color seen by a normal human eye from a distance large enough not to distinguish the individual pixels, as represented in the right square. The next note gives more detail.

composition (shown by the three color bars) to the visually perceived color (shown in the uniform square). If you wish to understand this connection better, please read the following note.

SPECIFICALLY, HOW IS FULL WHITE PRODUCED FROM THE THREE DARKER COLORS RED, GREEN AND BLUE? The first line of Figure 6-11 reminds us how we overlapped red, green and blue circles to form white in Figure 2-13: in the RGB system, we get white only where the three colors actually overlap. We can achieve the same white by overlapping the three color bars R, G and B

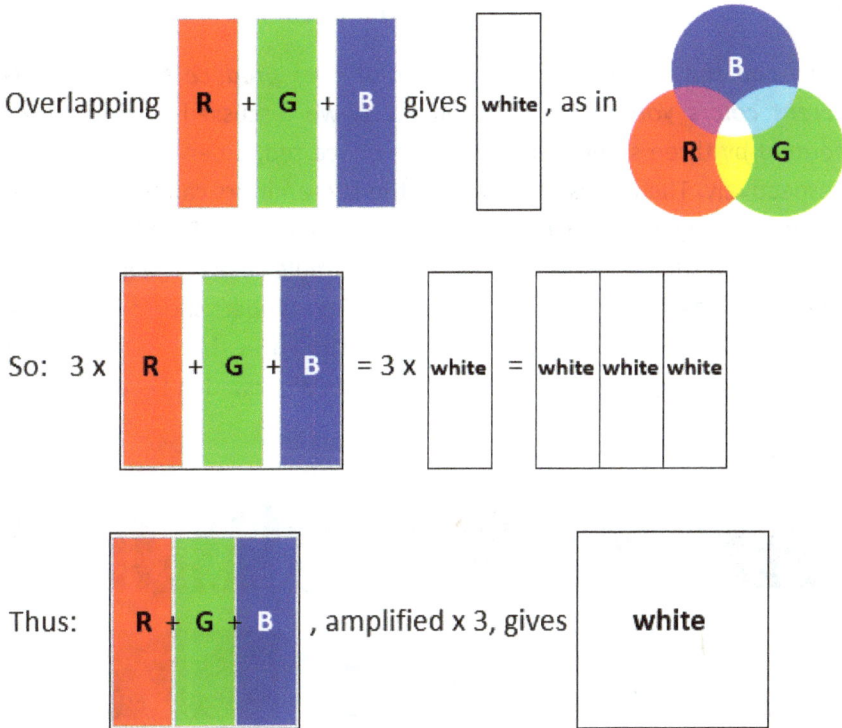

Overlapping R + G + B gives white , as in

So: 3 x R + G + B = 3 x white = white white white

Thus: R + G + B , amplified x 3, gives **white**

Figure 6-11: The top line reminds us that by overlapping full-intensity red, green and blue, we get normal white, as we already saw with RGB color circles in Figure 2-13. This fills the equivalent of one third of a pixel with white. The second line multiplies the foregoing result by 3 to fill the remaining two thirds of the pixel with white light. The last line summarizes by stating that amplifying the full-intensity red + green + blue light by a factor 3, we can fill one emitter pixel with white light.

(Continued)

(Continued)

(these represent three sub-emitters giving the same maximum light output, at level 255 in terms of RGB intensities): so the bar marked "white" concentrates the intensities of the three bars marked R, G and B. In the second line of Figure 6-11, we simply multiply by three the intensities from the first line, so that we get three white bars that can be placed side-by-side to fill a square pixel that is completely white. The third line summarizes these steps: it points out that, by amplifying three-fold the light emitted from the three sub-emitters, we produce enough light to fill a square pixel with white light.

6.2.2   *Creating nearly 17 million colors*

How can the red, green and blue color bars produce the many different colors you see? In Section 6.2.1, we discussed emitter pixels formed by three sub-emitters which produce red, green and blue light, respectively. The intensity of each of the sub-emitters can be controlled independently by the RGB intensity levels, from 0 to 255. In Figure 6-12, I show such square pixels producing some simple colors. The intensity of the color in each color bar is given by the RGB numbers shown just above the pixels. The eye perceives the colors shown in the second row of boxes (after multiplying the intensity by 3, as discussed in Section 6.2.1).

Normal vision: From RGB color to perceived color

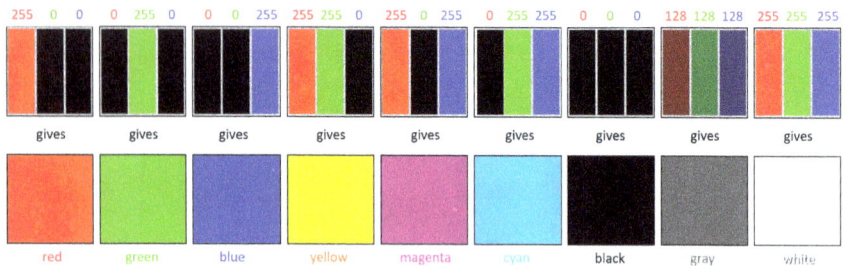

Figure 6-12: Sketched are RGB emitter pixels in the top row and the corresponding color perceived by the normal human eye in the bottom row. The emitter pixels contain red, green and blue sub-emitters; their emission intensities are given by the RGB numbers shown above each sub-emitter, with a maximum intensity of 255. The perceived colors are those seen from a distance such that the eye does not discern the individual pixels. (The exact correspondence between the emitter pixels and the perceived colors is discussed in Section 6.2.1.)

You can clearly see many of these colors in Figure 6-8, where the sub-emitters are photographed on a display. In Figure 6-12, a mid-gray pixel is shown (the intensity level 128 is about half of 255); for comparison, Figure 6-9 includes emitter pixels with a variety of gray levels, ranging from black inside the letters to white outside the letters.

More complex colors are shown in Figure 6-13. You can find on the internet long lists of colors and their corresponding RGB intensities. We will describe in Section 6.2.3 how to use such RGB intensities to color images and text.

Normal vision: From RGB color to perceived color

Figure 6-13: As Figure 6-12, but showing more complex colors.

DOES THE "GOLD" COLOR SURPRISE YOU? In Figure 6-13, the leftmost color is labeled "gold". Is this color your idea of gold, or does it look more like the yellow shown in Figure 6-12? If you look at the RGB intensities of these two colors, you will find them to be rather similar: (255,215,0) for gold *versus* (255,255,0) for yellow. In other words, gold has about 16% less green than does pure yellow: is this a satisfactory answer for the "special" appearance of gold? No! Gold looks special because it has a metallic appearance which gives it a metallic "shine". Without the "shine", a gold surface looks essentially like unexciting brownish yellow! Such shining shows up as intense reflections on many metals (unless they are rusted or painted over) and most glassy and crystalline objects: the shine is mainly due to the very smooth surface and is lost by scratching or rusting. The RGB colors can reproduce these shiny reflections in an image: indeed, a photograph or painting can record such reflections and produce the desired appearance. We also discussed "shine" in Section 4.2.

As explained in Section 2.1, the color names and exact color definitions are not universal: you may therefore produce your own varieties of salmon, turquoise, *etc.* on your own computer. You may even define your own primary colors different from red, green and blue, as long as they can be combined to produce all desired colors.

As we have already discussed, even white is not uniquely defined. Take a sheet of white paper around your home to compare it with "white" walls, furniture, bedsheets, clothes, crockery, ceramic tiles, television screens, computer displays, *etc.*: you will see many variants of white, such as creamy, bluish or grayish. As you turn or bend a sheet of paper, its own white will vary with the light direction falling on it. Also, white light varies as it is generated by different sources like the Sun, lamps and computer displays, because their spectral compositions differ. This is even the case between electronic displays from different manufacturers. As an example, the photograph in Figure 6-14 shows five color displays with the same white pages and letters in primary RGB colors (with no other lighting in the room): in this particular case, the top left TV display is more blue, the top right desktop display is more yellow, the two midsize laptop displays are more gray, while the small smartphone display is brighter but "warmer", namely a bit redder.

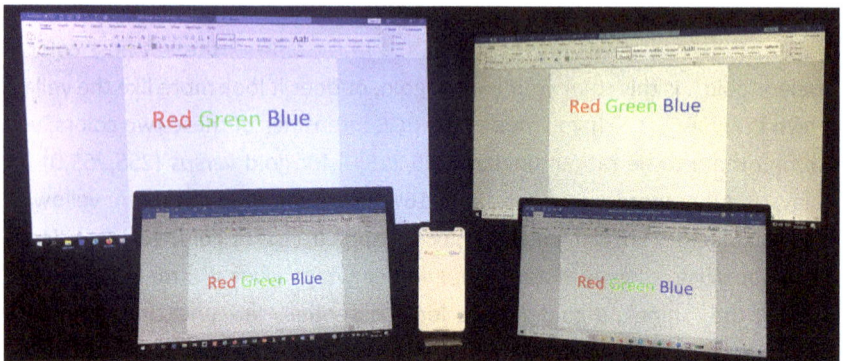

Figure 6-14: Single photograph of five electronic displays showing the same white page with the same primary red, green and blue colored text: at top left is a television monitor; at lower center is a smartphone (somewhat overexposed); the left and right lower displays are on laptops, while the upper right display is a separate desktop computer monitor.

HERE IS A SLIGHTLY MORE TECHNICAL QUESTION: ***Can we exactly reproduce the colors of the solar spectrum using RGB colors?*** The answer may come as a surprise: we cannot combine RGB colors to represent any of the colors of the solar spectrum! Consider the colors of the color triangles shown in Figure 2-5 and Figure 2-8. These colors are not pure solar spectral colors, as they are not present in the solar spectrum. In particular, my simulation of the solar spectrum in Figure 2-11 was made with RGB colors and therefore cannot be quite accurate.

You can imagine the solar spectrum to lie on a curved line that surrounds the color triangle of Figure 2-5: the red end of the solar spectrum would lie outside the red corner of the triangle; the green central part of the spectrum would form an arc around the green corner of the triangle and its blue end would lie close to the blue corner of the triangle. So the solar spectrum lies completely outside the color triangle. The solar spectrum is also shown in Figure 6-15; here the area outside the RGB color triangle (marked as **sRGB**) is expanded to show the location of the solar spectrum, which forms the horseshoe-shaped curve called "solar locus". This so-called **CIE 1931 color space chromaticity diagram** is used for technical applications, and is explained further in the caption of Figure 6-15. Terminology: a range of colors like the sRGB triangle or Rec. 2020 triangle in Figure 6-15 is often called a **gamut** in the color literature; the colored shape of this diagram is also a color gamut.

The colors in the strip outside the RGB triangle and inside the solar spectral line also cannot be reproduced with RGB colors.

What do the colors that are missing from the RGB color triangle actually look like? Essentially, they are more saturated and therefore more "colorful" or "vivid" versions of the colors that we see in the color triangle of Figure 2-5: so you can think of them as being less foggy than the colors in the color triangle. Remember that the center point of the color triangle is white/gray, and thus "totally foggy" and "colorless"; going outward from the triangle's center gradually reduces that fog by removing the gray component; at the edge of the RGB triangle there remains some fog, which can only be removed by going outside the triangle; once you reach the solar spectrum, you have "pure color with no fog", which is total saturation, as discussed in Section 4.1 for the HSL color system.

Using the RGB colors we can only provide "foggy" ("whiter") approximations to the solar spectral colors and the other colors outside the RGB triangle: for example, for an outside color, the nearest color inside the RGB triangle can

(Continued)

(*Continued*)

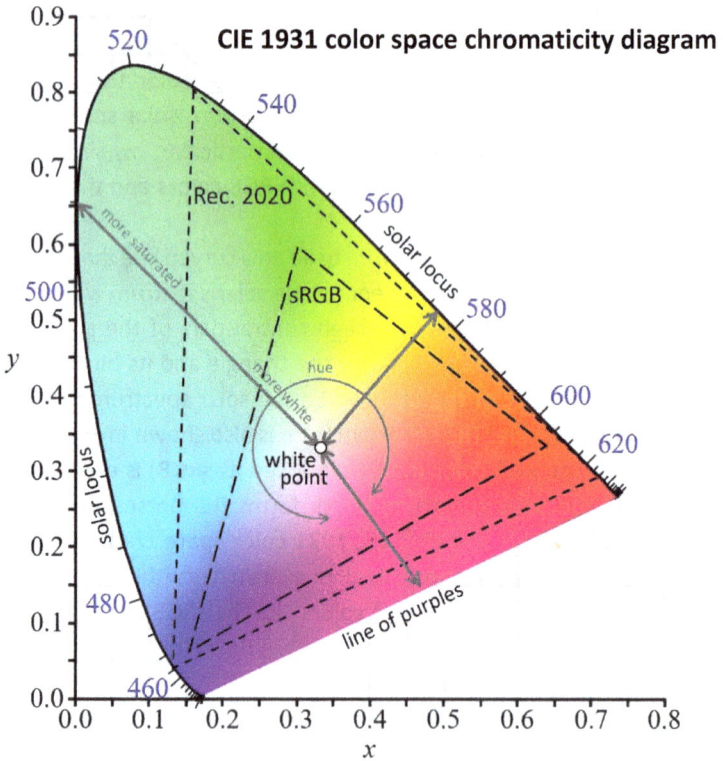

Figure 6-15: This diagram was developed in 1931 by the International Commission on Illumination (with French name Commission Internationale de l'Eclairage, hence CIE) to standardize colors in practical applications, especially for lighting, printing and electronic displays. The curve in the shape of a horseshoe represents the solar spectrum, called solar locus (the blue numbers along the solar locus give the wavelengths in nanometers; see also Section 8.2 and Figure 8-2). The inner triangle labelled sRGB is a specific example of RGB triangle, similar to the color triangle of Figure 2-5: sRGB is used by Microsoft and in many electronic displays and printers (other RGB triangles, such as Adobe RGB, are similar to sRGB, but have their red, green and blue corners in different positions in the CIE diagram). The outer triangle labelled Rec. 2020 is an expanded and therefore richer color triangle used with ultra-high definition television (UHDTV). The entire colored shape in the CIE diagram is the range of humanly visible colors. The solar locus consists of all pure (single-wavelength) colors that are present in the solar spectrum and visible to the normal human eye, extending from red at bottom right, through green at top, to blue at bottom left: these are the most saturated ("colorful", "vivid", "pure")

Caption continued on facing page

be chosen. Fortunately, the RGB triangle comes close enough to those colors, so that we don't normally miss them when we watch digital photographs, computer monitors, television displays or movie theatre screens.

However, since 2012 a new set of colors called **Rec. 2020** has been replacing the older RGB (and intermediate **sRGB**) color sets, especially for **ultra-high definition television (UHDTV)**; Rec. 2020 stands for "Recommendation 2020" and was further updated to Rec. 2100 in 2016. Essentially, Rec. 2020 and Rec. 2100 expand the color triangle of Figure 2-5 to include more colors that we can perceive, as shown in Figure 6-15. The result is television images with more saturated, "colorful" and vivid colors. Nonetheless, these additional colors still do not include the solar spectral colors.

Most television manufacturers are adopting the new richer color scheme. The Rec. 2020 scheme can also enrich electronic painting (e-painting or digital painting), when displayed on UHDTV screens.

We conclude: Our eyes can see all visible colors of the solar spectrum, but we cannot reproduce these colors exactly with the RGB color scheme; expanded schemes like Rec. 2020 add colors that are more saturated, more vivid. On the other hand, the primary colors red, green and blue used in the RGB scheme can be produced by combining solar spectral colors: these are however not easy to get except from sunlight or some lasers.

(Continued)

Figure 6-15 on facing page

colors possible. The straight "line of purples" does not exist in the solar spectrum, but its colors (mixtures of red and blue) are visible to the human eye. The central point labelled "white point" is the most "colorless" or most "foggy" color (it also represents gray). Color saturation increases in all directions from the white point to the solar locus and line of purples. Moving around the white point means cycling through the hues, from red at right, *via* green at top, to blue at bottom left, and on to red again (just like the circle of complementary colors of Figure 2-17. The axes *x* and *y* very roughly indicate the amounts of red and green, respectively, sensed by a normal human eye: basically, a higher *x* means redder, while a higher *y* means greener (most blues actually exist in a third dimension *z*, not drawn in this diagram). Note that this figure is produced with sRGB technology for this book and for your display: therefore, the colors seen outside the sRGB triangle are not correctly reproduced and are only approximations of the real colors seen by the human eye, as is also true of the solar spectra shown in Section 2.5.

(This image is based on the CIE 1931 xy color space diagram of Wikipedia at https:// commons.wikimedia.org/wiki/File:CIExy1931.svg, which is in the public domain, with additions by this author.)

(Continued)

Such colors, however, are available with the CMY system used for printing and painting. The reason is that CMY is a <u>subtractive</u> process. Consider white sunlight or red laser light falling on a piece of paper onto which an image or text has been printed with the CMY process. The colors that you see are those of the reflected light that has <u>not</u> been absorbed by the ink or paint. So you will see all the solar light or laser light that is reflected, including such colors that fall outside the RGB triangle.

(Strictly speaking, if we assume that the CMY system is derived from the RGB system, as in "cyan = green + blue", then the CMY system is also limited due to the limitations of the RGB system. However, if we look at CMY as absorbing certain kinds of light, then we are not limited by the RGB system's constraints: this is the point of view of the preceding paragraph.)

In summary: The RGB scheme does not include all possible visible colors, while all possible visible colors can be produced by combining solar spectral colors.

This conclusion also means that the colors of most photographs of the solar spectrum that you find on the web (and in this book) are only approximately correct. The reason is that the web (and this book) use the RGB color scheme, which cannot correctly produce the colors of the solar spectrum. More generally, photographs on the web are only approximately faithful in terms of their colors.

6.2.3 *How can we actually create the 16,777,216 different RGB colors?*

It is quite easy to make any of the nearly 17 million RGB colors on your computer, for example in today's Microsoft Word, PowerPoint or Excel, which use the same approach described below. I shall assume that you use Microsoft Word, but you can use PowerPoint or Excel as well. First, let's color the single word "color" by following these instructions:

- Type "color" or any other word into a Word document and select this word.
- A popup window appears, similar to this one (Figure 6-16):

Figure 6-16: Microsoft Word's popup box for selected text.

• Click on the downward arrowhead to the right of the red-underlined letter A (you can do this also on Word's ribbon in the Home|Font box). You should see a popup window like this one (Figure 6-17):

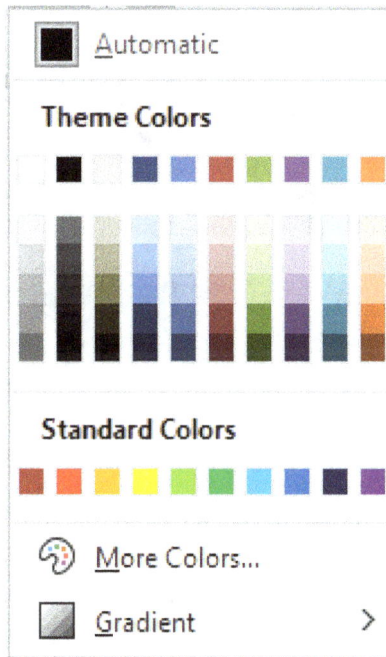

Figure 6-17: Word's first color popup box.

• This window gives you only 70 possible colors: you can click on any one, for example the light blue in the bottom row of colors, and you will see your word change to your selected "color". If you want another color that is not among the 70, go to the next step.

- To find more color options than the 70 shown in the last popup window, click on More Colors... You should now see a popup window like this one (Figure 6-18):

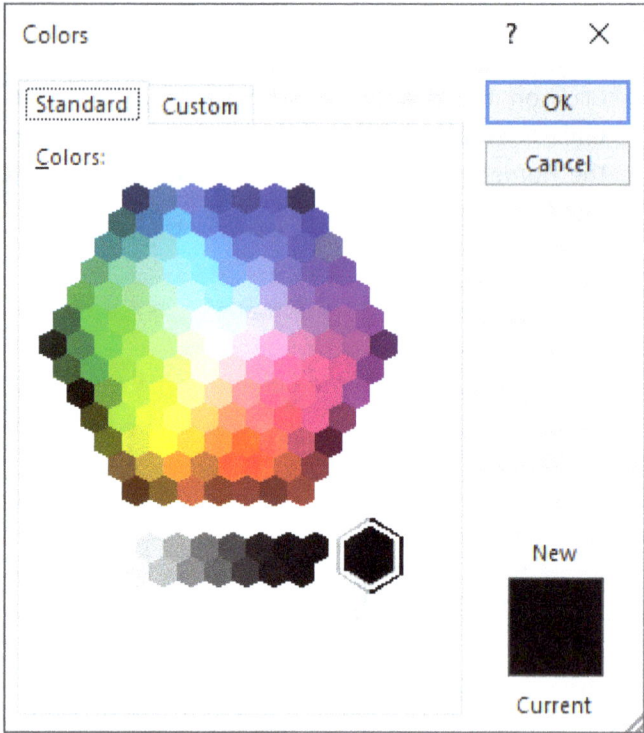

Figure 6-18: Word's second color popup box.

- This new window gives you 142 color choices, simply by clicking on one and then on OK. For example, the bottom right red/ brown choice gives you a brownish red "color". If you want another color that is not among these 142, go to the next step.
- For further choices, click on Custom in the last popup window, leading to the next window (Figure 6-19) (if you selected a non-black color before, the + sign in the rectangular color palette and the intensity slider at right will point at it):

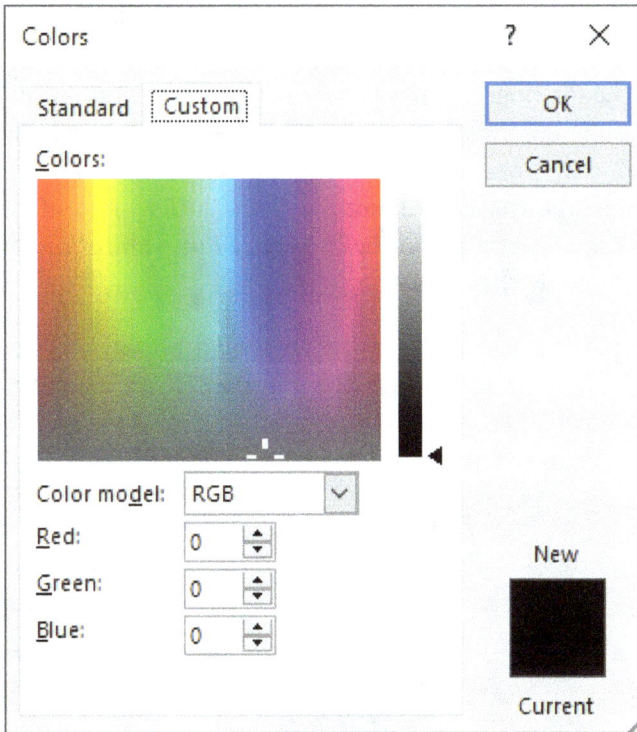

Figure 6-19: Word's third color popup box, for RGB color selection.

Here you see a palette of colors at left and an intensity slider to its right. You can drag the white + sign (partly hidden at bottom in this popup view) across the palette to choose any color seen there; you can also change its intensity with the slider at right (you will see your choice of color being continuously updated under the word "New", in comparison with the "Current" color that will be replaced by the "New" color). By dragging the + sign and the intensity slider, you can create any of the 16,777,216 colors allowed by the RGB system. If you need finer control of the color, you may use the RGB or HSL systems described below. If you are satisfied with the color, click OK to apply the new color to the word that you are coloring.

You do not get very fine control of the color by dragging the + sign and the intensity slider in the popup window shown above (Figure 6-19). If you want finer control, you can use two alternative methods by specifying numerically the intensities of red, green and blue in the RGB system, or the hue, saturation and luminance in the HSL system.

Since the **HSL color system** is more intuitive, let's first use it here: *How do we specify a color by its precise hue, saturation and luminance values?* Follow these steps:

- In the last popup window shown above, you see the words "Color model: RGB". Use the drop-down menu to select "HSL" instead of "RGB". The bottom left part of the popup window will change to the following (Figure 6-20):

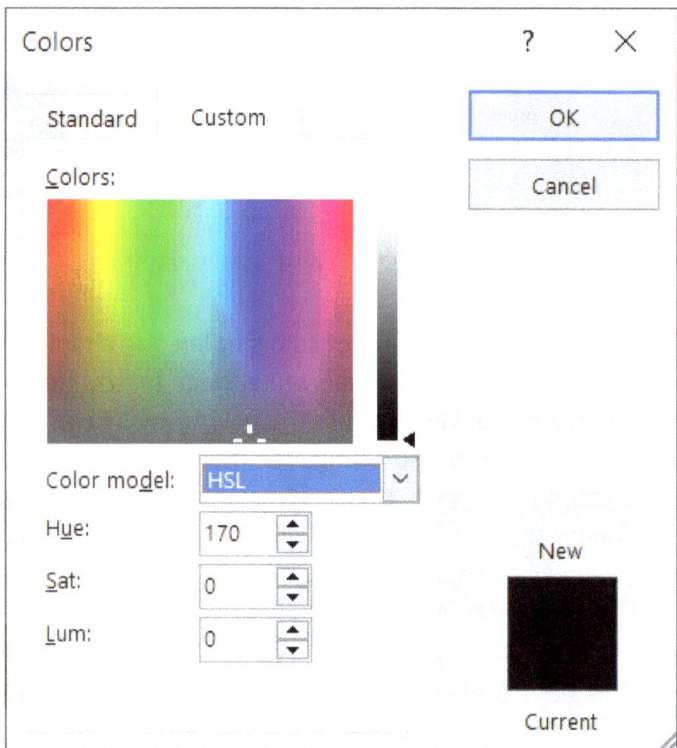

Figure 6-20: Word's third color popup box, for HSL color selection.

- Now you see lines with numbers for Hue, Sat(uration) and Lum(inance): the values are any whole numbers between 0 and 255. We have already described hue, saturation and luminance in <u>Section 4.1</u>. It is easy to see their meaning in this popup window. The hue varies in the rectangular palette, horizontally from left to right, from red through yellow, green, cyan, blue and magenta back to red (with value 0 at left to 255 at right); the value 170 shown corresponds to a bluish hue. The saturation varies in the rectangular palette, vertically from bottom to top, from gray ("colorless") to gray-free ("colorful"), with value 0 at bottom to 255 at top. The luminance is shown by the vertical black-to-white intensity bar, which varies from 0 at bottom to 255 at top. You can type in new values for these numbers, or increase/decrease them with the up/down arrows.

- Then click OK to apply your selected color.

How do we specify a color by its precise red, green and blue values? If you prefer the **RGB color system** to the HSL system, follow these steps.

- Go to the earlier popup window showing "Color model: RGB" and three numbers after Red, Green and Blue (Figure 6-19). These numbers can be any whole numbers between 0 (meaning zero intensity) and 255 (meaning full intensity); for example, "Red: 255" means full intensity of red, while "Red: 128" means half of full intensity of red, which is a darker red. These numbers give you the 256 intensities of the three primary colors red, green and blue that we discussed before (the number 0 is also counted, giving 255 + 1 = 256).

- You can type in new values for these numbers, or increase/decrease them with the up/down arrows. If you choose "Red: 255" and "Green: 255", leaving "Blue: 0", you get yellow. White is given by entering three times the number 255. To produce gray, type in three equal numbers between 0 and 255, for instance three times 128: this intensity gives you a gray with about half the intensity of white.

- Click OK to apply your selected color.

The RGB and HSL methods discussed above for selecting among the 16,777,216 available colors are not only valid in Microsoft Word: they are also available, for example, in Microsoft PowerPoint for making color drawings and in Microsoft Excel for coloring spreadsheets. One interesting option is "gradient" colors (see at bottom of Figure 6-17): this option allows you to make a smooth color variation like this from red to blue across text, image or spreadsheet (that is how I made the simulated solar spectrum of Figure 2-11).

Here is one application of color changes: ***How does color change over time? Can we distinguish older from newer documents by their color?*** Inks and paints tend to fade over time: color fading is a loss of colorfulness, an increase of "fogginess", and thus a loss of saturation. For example, the next two colored lines are the titles of two copies of Mr. John Doe's will: which one do you think looks more recent and therefore legally valid? Does this also allow detecting faked wills? Or forging ancient-looking documents?

Last Will and Testament of John Doe
Last Will and Testament of John Doe

6.3   Sources of light

Where does light come from? We have mentioned a few **light sources**, such as the Sun and other stars, lamps, and emitter pixels in electronic displays. We will here also mention other sources of light, such as, surprisingly, the human body itself! In addition, we will sketch the physics that generates light in those sources.

Let's start with the **sunlight** that we receive directly from the Sun, or indirectly by scattering through the clouds or reflection from water or glass (we even receive some sunlight by reflection from the Moon). Sunlight is by far our largest source of light, but also our main source of energy, including through photosynthesis that grows plants and thus provides us with food, coal, oil and gas; the Sun also supplies solar energy (both as heat and as electricity), and it causes weather and therefore rivers that provide hydroelectric power. Sunlight comes from a gigantic **nuclear reaction** that rages permanently within the Sun: it is a huge **hydrogen bomb** that explodes non-stop (already since about

4.5 billion years and likely for as much time into the future). The Sun contains mostly hydrogen (often abbreviated to H, the same atom found in water, H_2O). Basically, in the Sun as well as in a hydrogen bomb, the nuclei of hydrogen atoms combine in a reaction that forms the nuclei of another kind of atom, helium (abbreviated to He, which is often used to fill balloons). Importantly, according to Einstein's theory of special relativity, that combination of hydrogen to helium releases a large amount of energy in the form of light and other electromagnetic radiation (as well as other energetic particles). The emitted light is what we see coming from the Sun to Earth. It comes with all possible colors that form the solar spectrum (shown in Figure 2-12). It also includes invisible electromagnetic radiation, such as radio waves, microwaves, infrared light, ultraviolet light and x-rays.

Light can also be generated by chemical reactions, most obviously seen in a **chemical explosion** or **fire**, but also in **bioluminescence**, which is the generation of light by living beings such as fireflies at night and fishes that live in deep dark waters. In some chemical reactions within those organisms, energy is liberated and converted into light. The color of this light depends on the particular chemicals involved in the reaction.

Similar to biological light emission, solid materials can also produce light by **electroluminescence**: in particular, this is the case of the light emitters in **electronic displays**. Here an electric current caused by an electric potential provides energy that is converted to light. The color of the light depends on the material used: thus, different materials are chosen to create red *versus* green *versus* blue light.

Lasers use a similar method to create light with special materials (the name laser is abbreviated from "light amplification by stimulated emission of radiation"). In addition, however, lasers have a special kind of "resonance chamber" that favors one wavelength (and therefore one solar color); in practice, the wavelength of any given laser light spans about 1 nanometer, which is about 0.3% of the width of the visible spectrum shown in Figures 2-9 and 2-11 or about 17% of one slice in Figure 2-12 (for comparison, red light from electronic displays has a width of about 60 nanometers, while blue light is about 30 nanometer wide, green being in between). Importantly and uniquely, a laser also arranges all the light waves it produces to be "in phase", so that all the wave tops coincide and reinforce each other, giving very intense light.

Electrical generation of light also occurs in **lightning**, in a spectacular and thunderous way: friction in clouds charges those clouds electrically (like the rubbing of plastic against cloth), until a long spark flashes between clouds or between a cloud and the soil. In a spark, and therefore also in lightning, an electric current rips electrons out of atoms in a gas (like air). This is called a plasma in physics: in a plasma, reactions between atoms and electrons can emit large amounts of light (and noise). The same physical principle is tamed in a fluorescent lamp, often called **fluorescent tube**: it contains a permanent, gentle and quiet lightning stroke that emits constant ultraviolet light, helped by a phosphor coating inside the tube to make it visible, in a color that depends on the gas and coating used. The same principle is also used in older **plasma TVs**: there each emitter pixel contains a small steady lightning stroke emitting light of the appropriate color.

Another approach to producing light is **fluorescence** or **phosphorescence** in a solid material, as opposed to a gas. In this case a material is exposed to one color of light, after which it emits light of another color. With fluorescent materials, this happens almost instantaneously, so we have essentially an immediate conversion from one color to another; for example, to make forgery difficult, banknotes may use ink that becomes visible only when exposed to invisible ultraviolet light. With phosphorescent materials there is a time delay up to several hours; many older watches use this principle to show time in darkness.

Warning: some colors are commonly but incorrectly called "**fluorescent**" even though they do not involve fluorescence; they just look fluorescent. In particular, some clothes are colored "fluorescent" without having any fluorescent material in them. Such colors are so vivid (bright and saturated) that they seem too intense to be due to reflection alone; it's as if these colors came from a fluorescent material and would shine by themselves even in the dark.

Light is also spontaneously produced by any warm object. This is most obvious with electric stove elements used for cooking: those elements glow red when hot. When heated to higher and higher temperatures, all materials glow red, then blue and even white, as blacksmiths and glassblowers know very well (however, many materials melt or burn well before they even glow red, such as plastics and cloth!). Another example is the old incandescent light bulb: it has a filament that is heated by an electrical current and thus glows brightly; the heating of the lamp filament and of the electric stove element is due to electrical resistance. This radiation of light by warm objects is called **black-body radiation** in physics. The words "black body" are somewhat confusing in this name, since a body that radiates light is not truly "black"; the rationale for "black-body radiation" is that even a body that reflects no light at all (which therefore may be called "black") can actually spontaneously emit light. In fact, your body always and constantly emits black-body radiation; you can sense this radiation if you place your palm close to your cheek: you feel heat coming from your cheek; however, this heat is essentially invisible as it comes mostly in the form of infrared radiation. Another good example is a circulating-water radiator that heats a home, even without glowing red hot. Actually, any warm object emits light, although that light may be largely invisible.

In summary: **There are many physical, chemical and biological ways to generate light.** The Sun is our most abundant and cheap source of light, but it is not necessarily the easiest to use. There are many other sources that are more convenient if not cheaper, many of which are even available at the flick of a switch, the push of a button or the strike of a match.

6.4 What have we learned in this Chapter?

Printing on paper in color commonly lays down ink as tiny CMY-colored dots with varying sizes or density. Electronic displays emit RGB-colored light using emitters in each small pixel. To minimize the recharging of smartphones and computer pads, we should display text as white on black, not as black on white.

A computer display or television screen does not simply show the original pixels of the digital photograph; the image must first be converted from the grid of detector pixels to the grid of emitter pixels. The three sub-emitters of each emitter pixel correspond to the three basic RGB colors: together they can produce any of the 16,777,216 different RGB colors.

Our eyes can see all visible colors of the solar spectrum, but we cannot reproduce these colors exactly with the RGB color scheme. The RGB scheme does not include all possible visible colors, while all possible visible colors can be produced by combining solar spectral colors. More recent schemes, like Rec. 2020 for ultra-high definition television (UHDTV), add more saturated, "colorful" and vivid colors.

Many physical, chemical and biological sources of light are available, ranging from relatively inflexible sunlight to convenient emitter pixels.

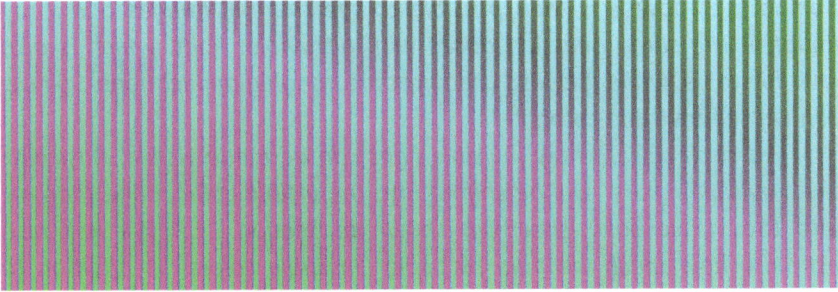

7

Seeing Invisible Colors

We have all seen x-ray images used for medical purposes and for checking baggage for dangerous contents. You may also have seen thermal images used to check people's temperature. Figure 7-1 gives two examples: these are made with infrared light and x-rays, respectively.

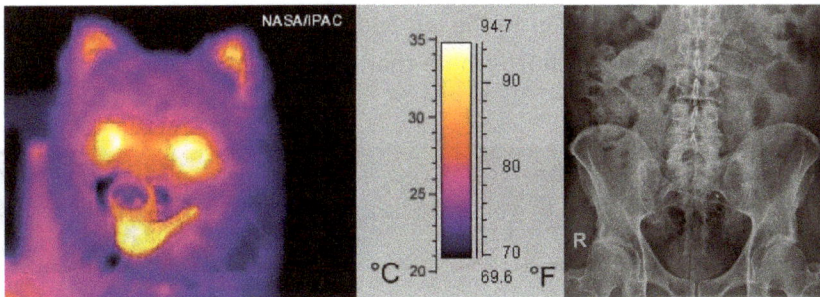

Figure 7-1: At left, an infrared image shows the high temperatures in the mouth, eyes and ears of a dog, while its nose and fur remain cool; the temperature scale in the center explains the coloring. (*Source*: NASA.) At right, an x-ray image shows my kidney stone as a small white dot, above and to the right of the center.

*However, neither of these kinds of light is actually visible to the naked human eye: how is it then possible to see thermal images and x-ray images? The same question arises with ultraviolet light, which is also invisible to us, while it is also used in some imaging applications. So how can we see **invisible colors**?*

*We shall discuss how invisible light can be made visible. Basically, the "invisible colors" will be shifted into the **visible color** range (namely into the visible solar spectrum shown in <u>Section 2.5</u>). We will also estimate how much we gain by using "invisible colors" in addition to visible colors: is it really worth the effort?*

— ·}} {{· —

7.1 The electromagnetic spectrum

To address the interesting challenge of seeing "invisible light", we need to consider again the **solar spectrum** and how it relates to **invisible light**. The center of Figure 7-2 shows the solar spectrum of Figure 2-11. As mentioned in Section 2.5, this solar spectrum is only that part of the entire **electromagnetic spectrum** which is visible to humans. The entire spectrum additionally includes, to the left of visible red: **infrared (IR) light**, **microwaves**, **radio waves**, *etc*. And to the right of visible blue: **ultraviolet (UV) light**, **x-rays**, **gamma rays**, *etc*. Only some of these are shown in the figure. The visible portion is actually a tiny part of the entire spectrum: while it occupies the center of Figure 7-2, the invisible parts extend much farther to the left and to the right than shown; those parts are drawn much shorter here than they are in reality.

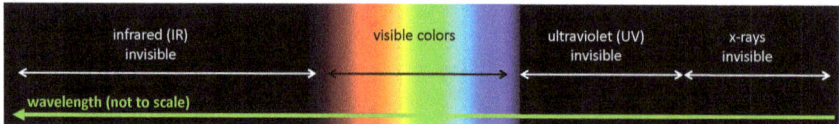

Figure 7-2: The visible part of the electromagnetic spectrum fits between the infrared and ultraviolet parts, drawn black here since they are invisible. The visible part is much narrower than the infrared and ultraviolet parts, which are not drawn to scale here (the wavelength increases to the left, the frequency to the right, but they are not drawn proportionally).

How can we see invisible light? The short answer is: We can see invisible light by "moving" invisible light into the visible range. There are several ways of doing this: we will describe four approaches.

7.2 Shifting invisible colors to visible colors: Thermal imaging

In a first approach, common in **thermal imaging (temperature imaging)**, we can literally shift a range of invisible colors of the spectrum into the range of visible colors. This is the case of the dog's infrared picture in Figure 7-1. You may have seen this approach used at checkpoints to measure people's temperature in order to check for fever.

The principle is illustrated in Figure 7-3. An infrared detector is used that is sensitive to the range which is shown as an orange box (the orange curve there shows an example of the infrared intensity coming from one pixel in an image, at different wavelengths).

The detected infrared intensities are then transferred (shifted) to the range of visible light, as shown by a white box; the box may be stretched or contracted to best fit the visible range: the intensities measured in the infrared range are now given to different colors in the visible range, as shown schematically by the white curve. Thereby, each pixel in the final visible image is given a color that mimics the corresponding invisible "color" of the infrared image.

Physically, this shifting from infrared to visible can be done in two ways. One way is to use infrared-sensitive film as detector (sometimes

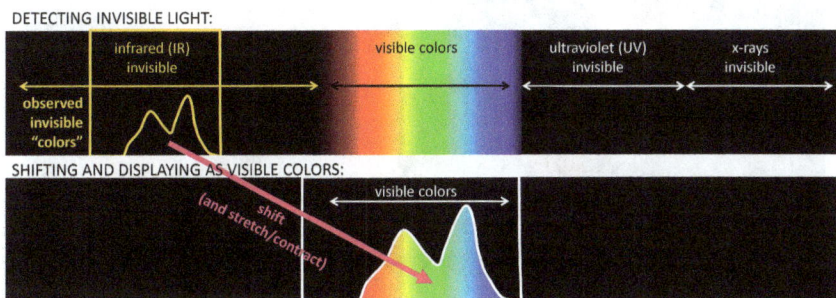

Figure 7-3: Infrared light is detected in the range indicated by the orange box at top left. It is shifted (and stretched or contracted as needed) into the visible range at bottom center: white box.

together with filters to eliminate non-infrared light). As with normal chemical photography, the film is then processed and results in a visible image: the resulting chemicals in the film are chosen to have various visible colors. Another way is electronic: a digital camera with special sensors (and filters if needed) records infrared light; the recorded information is then converted by computation to red, green and blue signals in the visible range for normal visible display.

Note that the range in the visible portion does not have to cover all colors from red through green to blue. The dog's IR picture in Figure 7-1 uses only some colors, from black to blue, magenta, red, orange, yellow and white, bypassing green (the magenta is a shortcut from blue to red in the color triangle of Figure 2-5; a more precise illustration of this color sequence is found in the color cube of Figure 4-2: the sequence starts at the black corner and follows the cube edges to the corners colored blue, magenta, red, yellow and white). Many other choices of color sequences are also possible, especially with digital images.

The principle behind thermal imaging with infrared light is that the wavelength of the infrared light emitted by the subject (dog, human, building, *etc.*) depends in a direct fashion on its temperature, so that the color ends up telling the temperature.

Another example of thermal imaging with infrared light is given in Figure 7-4. This image maps the **temperature** of the land surface of the

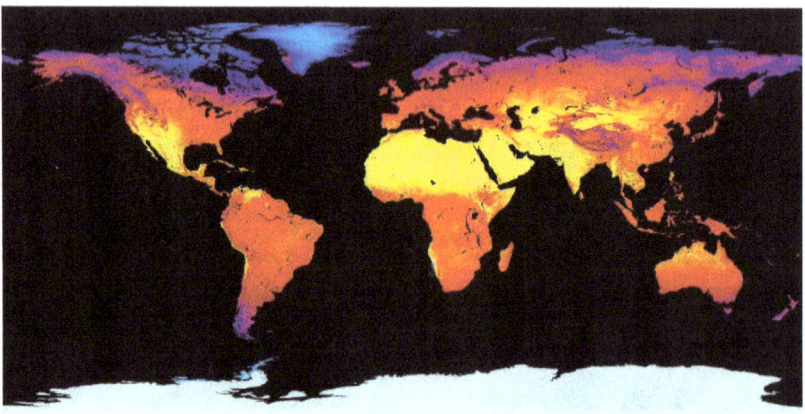

Figure 7-4: Thermal infrared map of the land surface temperature of the Earth, as measured by satellite. Here blue indicates cold temperatures, red indicates intermediate temperatures and yellow indicates warm temperatures. (*Source*: NASA)

Earth, as measured by a satellite. Such an image is a combination of many individual satellite pictures, digitally processed.

A different kind of infrared image is shown in Figure 7-5, which compares a photograph of a tree taken with a normal film to a photograph of the same tree taken with an **infrared film**. With this particular film, invisible infrared becomes visible red, visible red becomes visible green, and visible green becomes visible blue; visible blue becomes black because it was filtered out separately. In general, the exact colors vary with the chemical composition of the film, depending on what part of the infrared spectrum is detected and onto which part of the visible spectrum it is shifted. By contrast, we often see in other infrared images that green vegetation turns red; this is often the case with infrared mapping, which tends to show forests and fields in red, for example.

Figure 7-5: Tree photographed with normal "visible light" at left, and with infrared light at right. (*Source*: Dschwen — Own work, CC BY-SA 2.5, for UV picture[1] and for visible-light picture.[2])

7.3 Shifting invisible colors to visible gray: X-rays

A second approach is used in **x-ray imaging**. Here, a range of x-rays, shown at upper right in Figure 7-6, is shifted to the visible range, but as a simple gray color: darker gray for more x-rays. Remember that equal amounts of red, green and blue produce gray, so now the intensity is made equal across the whole visible range: thus, x-ray imaging converts

---

[1] https://commons.wikimedia.org/w/index.php?curid=707363.

[2] https://commons.wikimedia.org/w/index.php?curid=707365.

DETECTING INVISIBLE LIGHT:

| infrared (IR) invisible | visible colors | ultraviolet (UV) invisible | x-rays invisible |

observed invisible "colors"

SHIFTING AND DISPLAYING AS VISIBLE GRAY:

visible colors

make uniform and shift (and stretch/contract)

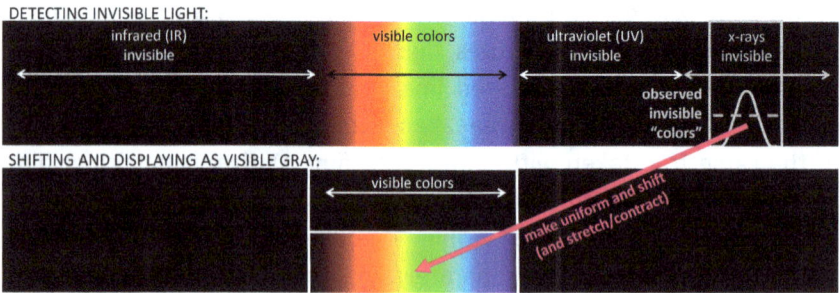

Figure 7-6: X-rays are detected in the range indicated by the light-gray box at top right. It is shifted into the visible range at bottom center (white box), after equalization of all colors: this produces a gray-scale color in each pixel of the image.

the x-ray intensity to gray levels for each pixel of the image. X-ray images can be made with a film that is most sensitive to x-rays, and which is then chemically processed like standard chemical film, such that regions with more x-rays become darker gray.

In x-ray imaging, the gray level reflects the density of matter in a body. X-rays easily penetrate soft living matter, but not bones: as a result, dense bones inside the body create darker shadows, as seen in Figure 7-1. As you may know, x-ray images are usually shown as negatives, so that **bones appear lighter gray than other living matter**; this is because we can better see variations in lighter gray than in darker gray.

The approach used with x-rays can also be used with **ultraviolet light** coming from the Sun or a lamp (normally, visible and infrared light must be filtered out). An ultraviolet portrait is shown in Figure 7-7. Here again, only a gray-scale image is produced. Both chemical films and digital detectors can be used.

Figure 7-7: Ultraviolet portrait using only UV light between the wavelengths of 335 and 365 nanometers. (*Source*: Spigget — Own work, CC BY-SA 3.0.[3])

[3] https://commons.wikimedia.org/w/index.php?curid=9023045.

Unlike x-rays, ultraviolet light does not penetrate skin deeply. Ultraviolet light highlights irregularities in skin more than does normal visible light: hence, skin looks much rougher in tone and texture with UV light than with visible light. **UV photography is in fact much used to detect irregularities in the surface of solid materials.**

7.4   Shifting invisible colors to visible RGB: Multispectral imaging

A third approach is often called **multispectral imaging**, because it combines light from different spectral portions of the electromagnetic spectrum. The principle is shown in Figure 7-8, while an example appears in Figure 7-9.

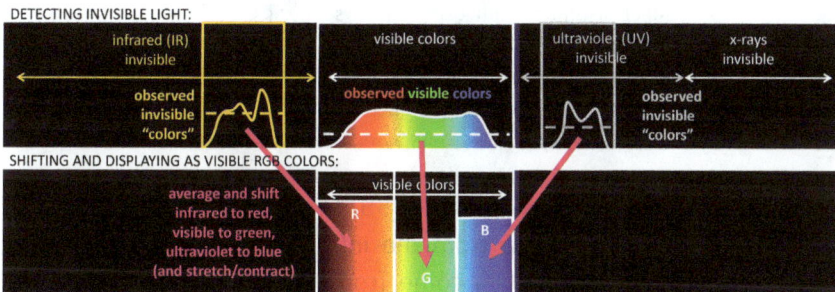

DETECTING INVISIBLE LIGHT:

| infrared (IR) invisible | visible colors | ultraviolet (UV) invisible | x-rays invisible |

observed invisible "colors" — observed **visible** colors — observed invisible "colors"

SHIFTING AND DISPLAYING AS VISIBLE RGB COLORS:

average and shift infrared to red, visible to green, ultraviolet to blue (and stretch/contract) — visible colors — R G B

Figure 7-8: Multispectral imaging uses two or more different parts of the full electromagnetic spectrum, here a part of the infrared spectrum (shown at left in orange), a part of the visible spectrum (shown in the center in white), and a part of the ultraviolet spectrum (shown at right in light gray). Here, they are respectively shifted into the red, green and blue parts of the visible spectrum, but only as three intensities that determine the red (R), green (G) and blue (B) components of a new visible RGB color for the pixel in question.

The general idea is to use light from two or more different parts of the spectrum; this may include some visible light as well. In the case shown in both Figure 7-8 and Figure 7-9, three parts of the spectrum are used: a part of the **infrared** spectrum, a part of the **visible** spectrum itself, and a part of the **ultraviolet** spectrum. The average intensity in each of these three spectral parts is given to red, green and blue components of a new visible RGB color of the resulting image pixel, as schematically shown in Figure 7-8.

Figure 7-9: Multispectral photograph using three spectral parts: infrared (giving the intensity of visible red), visible (giving the intensity of visible green) and ultraviolet (giving the intensity of visible blue). Specifically, infrared is mapped to the red channel (720–850nm), visible to the green channel (500–600nm) and ultraviolet (335–365nm) to the blue channel. (*Source*: Spigget — Own work, CC BY-SA 3.0.[4])

This approach is used in Figure 7-9. In this photograph, the predominantly red colors come from the infrared part of the spectrum, the predominantly blue colors come from the ultraviolet part of the spectrum, while the few greenish colors combine the visible colors of the original scene.

Another multispectral example is shown in Figure 7-10, which shows the presence and temperature of high clouds over a large part of the Pacific Ocean. The multispectral character of such images allows emphasizing very specific information, in this case atmospheric data useful to aviation.

Among countless other applications, multispectral maps similar to this can show the geographical distribution of agricultural crops and the status of their growth and harvesting, the various types of land use (urban, forested, agricultural, mining), or the level of pollution, but also winter snow cover important for spring river flow, ocean temperatures, ocean currents, sand and dust storms, *etc.*

---

[4] https://commons.wikimedia.org/w/index.php?curid=9023255.

Figure 7-10: Multispectral IR and visible image of temperature cloud tops over much of the Pacific Ocean, as measured by satellite. (*Source*: NOAA[5]; it is updated every 10 minutes.)

7.5   Making invisible colors visible by fluorescence or phosphorescence

In a fourth approach, it is possible to use **fluorescence** or **phosphorescence** to shift invisible ultraviolet light into the visible spectrum. We have described this process in Section 6.3. Briefly, the ultraviolet light is absorbed by a "fluorescent" material, which thereafter can emit visible light.

This effect is used, for example, in detecting **fake banknotes**. Under invisible ultraviolet light, fluorescent ink on a banknote will become visible. A fake banknote probably will not have that invisible ink and will then not light up under ultraviolet light.

7.6   What do we gain by using invisible light?

How useful is it to see "invisible" light? In the examples above we have noted important examples of applications that would be difficult

---

[5] https://www.star.nesdis.noaa.gov/GOES/fulldisk.php?sat=G17 — see "Sandwich".

without using invisible light: medicine, security, global land and ocean surface temperature, land use, weather, pollution, *etc*. This question is also important for insects foraging for food in flowers: they will be looking for specific flowers, so that any help in rapidly finding the right flowers is important. And indeed, insects (and many other animals) detect ultraviolet light which may be emitted by specific flowers, in addition to red, green and blue.

A related question is: **How much do we gain by using invisible light?** One way to think of this is to look at colorful flowers, as in Figure 2-1. With our three types of cones (red, green and blue), would seeing infrared or ultraviolet help a little bit or very much? In the following we will quantify in a simple way how much we gain by seeing additional primary colors: we will see that the number of combinations of primary colors increases rapidly.

Let's first ask what we gain by seeing more primary visible colors: that will help us before we jump into invisible colors.

In Figure 7-11 we see a part of the flower scene of Figure 2-1. If our eyes' cones only saw one primary color, for example red "R", we would see the **monochrome** view at the top of Figure 7-11. Everything has the same red color or hue; the only variation is the brightness of red (this is not actually quite the way colorblindness works, as we will discuss in Section 8.1).

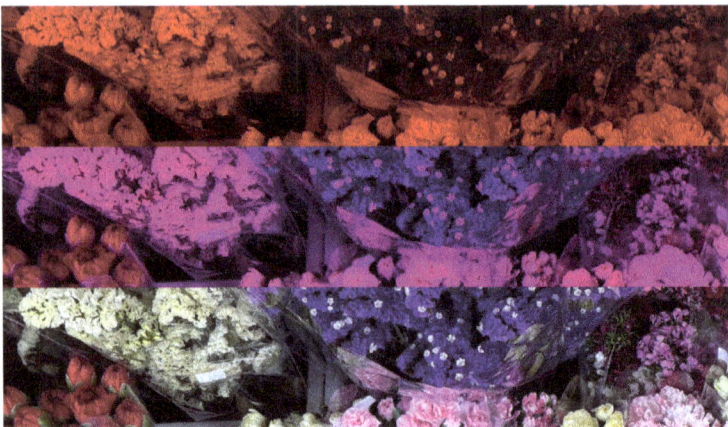

Figure 7-11: Comparison of flowers shown with 1 *versus* 2 *versus* 3 primary colors. Top: red only; middle: red and blue; bottom: red and green and blue. These images are extracted from Figure 2-1.

If our eyes' cones can also detect blue "B", we would see the middle frame, where both red and blue appear. Now we can also see pairwise combinations R&B (such as magenta or purple), as well as the brightness of the red component and of the blue component. Such combinations add a rich variety of "combined" or "mixed" colors that were not present with a single primary color. This variety of combinations allows distinguishing many different **hues**.

If our cones can also detect green "G", we can see the normal scene at the bottom of Figure 7-11 (with colors identical to those of Figure 2-1). Now red, green and blue appear; we also see pairwise combinations of these three primary colors: R&G, R&B and G&B. But we also get triple combinations of the three colors, such as white: R&G&B. The combinations create a large number of new colors, besides the three primary colors.

So, each time we add a new primary color, we not only gain one more basic color (and its various brightness levels), but we also get a large number of new combinations of the new primary color with the other primary colors. And the more primary colors we have, the more combinations become possible. This is also true when adding "invisible colors": they simply count as additional primary colors.

As an example, let's add an infrared primary color, which we will call "I", to the three normal primary colors (R, G and B). We thus add that infrared color with all its brightness levels, and also the pairwise combinations of that infrared color with red, green and blue individually: R&I, G&I and B&I. Furthermore, we add new triple combinations: R&G&I, R&B&I, G&B&I. But we also add a novel quadruple combination: R&G&B&I.

We can continue adding primary colors. We could add another infrared color (from another part of the invisible spectrum), or we could add an ultraviolet color from the ultraviolet portion of the spectrum: let's do the latter and call it "U". If we add the primary color U to R, G, B and I, we have five primary colors, together with all their possible combinations.

This process of adding primary colors, visible or invisible, rapidly adds many new combinations of colors. The following note will quantify these combinations.

WE CAN MAKE ALL THIS MORE PRECISE. READ HERE IF YOU ARE INTERESTED: We can indeed count the number of colors that we obtain when we add primary colors: this is done in the table shown in Figure 7-12. The table allows us to easily go to more than 3 primary colors, by adding further primary colors: these additional primary colors will be "invisible colors", since red, green and blue already cover the visible colors.

This table starts as we did above in Figure 7-11: in row 2 (entitled "1 primary color") the table shows the situation with one primary color, red (R). Red-only color vision, which looks like the top image in Figure 7-11, is a very rare example of color blindness. We assume 256 possible intensity levels, as in the usual RGB color system (see Section 4.1). With only one single color, we cannot create pairs of colors (column 3), nor triples ("trios", column 4), nor quadruples ("quads", column 5), nor quintuples ("quints", column 6).

Next, we add a second primary color, blue (B), shown in row 3 of the table (entitled "2 primary colors"). This situation corresponds to a slightly more common form of color blindness: green blindness. With two single primary colors, we can produce only one pair, R&B: this pair allows $256 \times 256 = 65{,}536$ color combinations, given the 256 intensity levels. The pair is shown as a colored bar connecting R and B, similar to the color bar of Figure 2-4: this bar shows the many hues that exist as combinations of red and blue.

Row 4 (entitled "3 primary colors") of the table adds a third primary color, green (G). This represents normal human vision. As we have seen in Section 4.3, three primary colors allow $256 \times 256 \times 256 = 16{,}777{,}216$ combinations. Pairwise combinations or R, G and B are shown in column 3 as three color bars forming a triangle: this is the edge of the color triangle of Figure 2-5. The interior of that color triangle is shown in column 4 as a trio: it contains all triplets that combine R, G and B.

Now we are ready to add a fourth primary color, in row 5 of the table (entitled "4 primary colors"). As discussed above, we now choose to add an infrared primary color (I). Since we cannot draw an invisible color, I artificially make it gray in this table. Four colors allow $256 \times 256 \times 256 \times 256 = 4{,}294{,}967{,}296$ or over 4 billion combinations: we already get a gigantic number of color combinations! There are six pairwise combinations or R, G, B and I, shown in column 3 as six color bars forming the edges of a 3D shape (you can see a 3D view of this shape in column 5). This 3D shape has four triangular faces, each of which contains triplets (trios) of color combinations, shown in column 4. The interior of the 3D shape can be imagined in column 5 (the interior cannot be drawn clearly): it contains all quadruplets (quads) of combinations of R, G, B and I.

No. of primary colors and no. of their combinations (assuming 256 possible intensity levels)	Single colors	Mixed pairs of colors (1D line)	Mixed trios of colors (2D interior of triangles)	Mixed quads of colors (3D interior of 4-corner polyhedron)	Mixed quints of colors (4D interior of 5-corner polyhedron)
1 primary color, 256 levels	1 single: e.g. red R	0	0	0	0
2 primary colors, 256² = 65,536 combinations	2 singles: e.g. R and blue B	1 pair:	0	0	0
3 primary colors, 256³ = 16,777,216 (~17 million) combinations	3 singles: red R, B and green G	3 pairs (lines like that above): 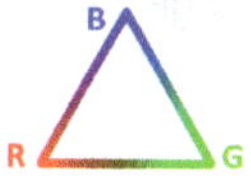	1 trio: 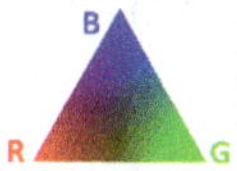	0	0
4 primary colors, 256⁴ = 4,294,967,296 (~4 billion) combinations	4 singles: R, B, G and "infrared" "I"	6 pairs: 	4 trios (triangles like that above):  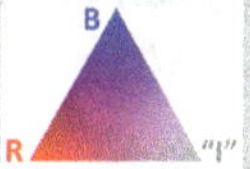 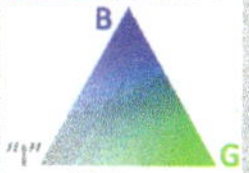 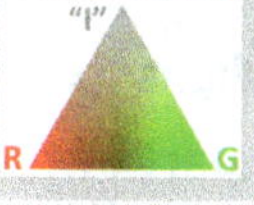	1 quad: 	0
5 primary colors, 256⁵ = 1,099,511,627,776 (~1 trillion) combinations	5 singles: R, B, G, "I" and "ultraviolet" "U"	10 pairs: 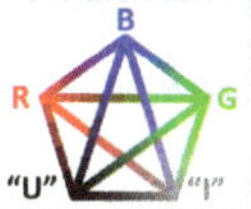	10 trios (triangles like those above): RGB, RGI, RGU, RIB, RUB, IGB, UGB, RIU, IGU, IUB	5 quads (polyhedrons like that above): RGBI, RGBU, RGIU, RBIU, GBIU	1 quint: 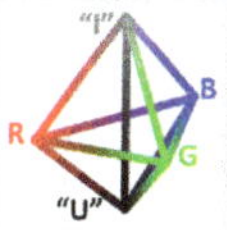

Figure 7-12: This table shows how many colors can be obtained by mixing 1, 2, 3, 4 or 5 primary colors. See the text for explanations. I call the 4th color "infrared" or "I", and the 5th color "ultraviolet" or "U"; I draw these two colors (artificially) as gray and black, although they are really invisible to normal humans. The table shows how many more mixing combinations become possible. The quads (5th column) can be seen as 4-cornered polyhedra in 3-dimensional space: all points inside them represent possible colors. Analogously, quints (6th column) can be seen as 5-cornered polyhedra in 4-dimensional space (this is impossible to draw in 2D, so the 5-polyhedron shown here is artificial): again, all points inside the polyhedron represent possible colors.

(*Continued*)

(Continued)

The story now continues likewise with a fifth primary color, in row 6 of the table (entitled "5 primary colors"). I chose this color to be invisible ultraviolet, called it "U" and drew it black. We get an astronomical number of combinations with the four other primary colors: over 1 trillion (which is over 1 million million). There are now 10 pairs (in column 3), 10 trios (in column 4, but not drawn here, as they would look somewhat like the trios above), 5 quads (in column 5) and one quintuplet (quint, in column 6); the quint is a shape in 4D space and thus cannot be drawn correctly in 2D or even 3D space.

In summary, we find that each additional primary color adds a large number of new color combinations with the preceding primary colors. More precisely, with 256 intensity levels at our disposal, each new primary color multiplies by 256 the number of possible color combinations: this is explosive growth, namely 25,600% each time! As mentioned in Section 5.1.4, the mantis shrimp detects as many as 12 to 16 primary colors: this produces a truly mind-boggling number of combinations; it is hard to imagine how the mantis shrimp uses all those colors! Even four or five primary colors offer an amazing richness of color combinations. That is why they can be so useful in applications such as imaging land use, crop status, weather, *etc.*

7.7   What have we learned in this Chapter?

We can see "invisible" light (x-rays, ultraviolet, infrared, *etc.*) by "shifting" it to the visible part of the solar spectrum. "Shifting" means recording the invisible light with suitable detectors, and displaying the resulting image with visible light; the image may be colorful (consisting of red, green, blue) or gray.

Important examples include: thermal imaging from infrared light; x-ray imaging in medicine; multispectral imaging of the Earth's surface by satellites; weather monitoring; detection of fake banknotes. Adding primary colors permits imaging more aspects of the scene, from a flower's colors to properties of the atmosphere, thanks to the existence of many added combinations of colors.

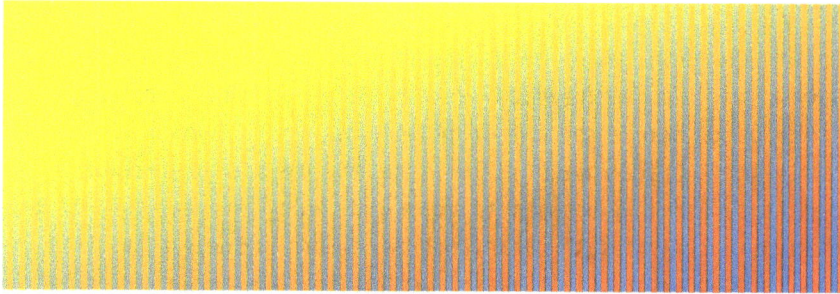

8

Color Blindness, Color Vision Deficiency and Normal Vision

*About 5% of humanity is "color-blind", better called "color-deficient". About half of those people are "red-green color-blind" and thus confuse red and green. We explore what "color-deficient" people see, and why. It comes down to the red, green and blue cones in our eye's retina, and to how the brain interprets the cones' signals. Also interesting is that we can at the same time learn a lot about "normal vision": we will find, for example, that even "pure red" light excites the "green" and "blue" cones to a considerable degree. We will also discuss the difficulty of communicating how we see **colors**, in particular but not only with color-blind people.*

—·❭❭ ❬❬·—

8.1 Surprises of color blindness and color vision deficiency

What is "color blindness"? Addressing this question will lead to several interesting surprises. These will also teach us useful insights into the mechanism of everyone's vision.

The first surprise concerning **color blindness** is that very few people are actually "color-blind", in the sense of not being able to see certain colors, like green. Only a few people are truly blind to one or more colors, such as blind to green (which we will call "green-blind"). Instead, most "color-blind" people see one or more colors weakly (which we will call, for example, "green-weak"). Therefore, the medical term **color vision deficiency** (often shortened to **color deficiency**) is more appropriate than color blindness; this term includes color blindness as an extreme case, but allows a wide range of weakness in color vision from nearly normal vision to full color blindness.

For convenience, we will use terms like "green-weak" and "green-blind"; however, we will rapidly discover that this terminology is really too simple.

How many people are affected? About 5% of humans have color deficiency, which is mostly inherited and incurable: this affects about 9% of males and 0.5% of females (this difference is related to the genetic inheritance path, which makes it much more likely for men to exhibit color deficiency). For comparison, about 0.5% of humans are totally blind, meaning that they see nothing, while about 4% are "visually impaired" and about 8% have some degree of "vision loss" (weak color deficiency does not qualify for these categories).

We see from the percentages indicated in Figure 8-1 that "green-weak" (more properly called **deuteranomaly**) is the most prevalent deficiency among humans: it affects about 2.7% of the population (or about 5% of males). Here the green cones react relatively weakly to green light. The "green-blind" deficiency afflicts about 0.6% of humans (or about 1.1% of males), similar to "red-weak" and "red-blind". These four groups together are often called **red-green color-blind**: many of them confuse reds, greens, browns and oranges; they may also confuse blues and purples.

Color-blind people can encounter job restrictions, because certain professions rely on good color discrimination, such as: airplane piloting, air traffic control, train driving, police, and firefighting.

What do color-deficient people see? Consider Figure 8-1: it illustrates roughly how different kinds of color-deficient people see the normal full range of colors (or hues as we also called them). The colors shown are those arranged around the **color triangle** of Figure 2-5 in

Approximate color perception:
normal vs. deficient
(with % of population affected)

normal vs. red-weak and red-blind

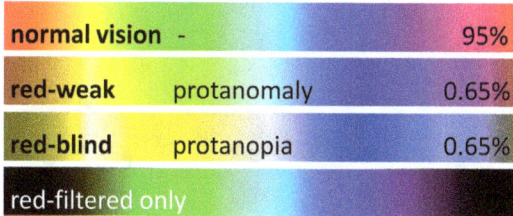

normal vision -		95%
red-weak	protanomaly	0.65%
red-blind	protanopia	0.65%
red-filtered only		

normal vs. green-weak and green-blind

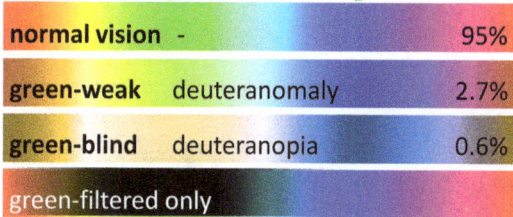

normal vision -		95%
green-weak	deuteranomaly	2.7%
green-blind	deuteranopia	0.6%
green-filtered only		

normal vs. blue-weak and blue-blind

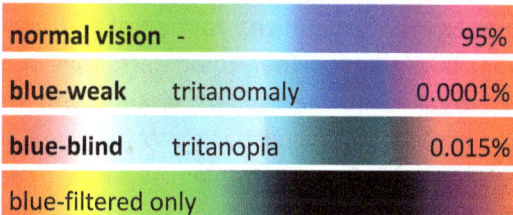

normal vision -		95%
blue-weak	tritanomaly	0.0001%
blue-blind	tritanopia	0.015%
blue-filtered only		

Figure 8-1: The most important types of inherited color deficiency are compared with normal vision for a wide range of colors, for three groups of people: red-weak and red-blind at top; green-weak and green-blind at center; and blue-weak and blue-blind at bottom. In each group, the top color bar shows normal vision: it includes all colors around the color triangle (Figure 2-5), from red through green and blue back to red. For each group, two conditions are named: color "-weak" (or "-anomaly") and color "-blind" (or "-anopia"); the colors are named "prot-" (meaning "first" in Greek, red being first), "deuter-" (second or green) and "trit-" (third or blue), hence: **protanomaly** and **protanopia**; **deuteranomaly** and **deuteranopia**; **tritanomaly** and **tritanopia**. These color bars show the colors that color-deficient people report seeing; they are quite variable from person to person, so these color bars are only representative examples. The fourth bar in each group of four shows a simple "filter model": it is obtained by covering the

Caption continued on next page

Figure 8-1 on previous page

normal-vision color bar with a partly transparent black layer to simulate the absence of detection of red or green or blue colors; it should be compared with the color bar right above it, labeled "-blind".

The percentages of the population affected by each type, shown at right, are derived from Wikipedia.[1] The corresponding color bars are in the public domain,[2] drawn by Nanobot and reorganized in this figure.

CAUTION FOR PEOPLE WITH COLOR DEFICIENCY: the color bars below those for normal vision are meant to be viewed only by people with normal vision; if people with a color deficiency look at these bars, they will add their own eyes' color deficiency to the color deficiency that is already built into these bars: the result will thus be doubly deficient and therefore not realistic. It is challenging to produce color bars that would look realistic to color-deficient people.

Section 2.3, and also displayed in Section 6.2.3 in a rectangular color palette. (A word of caution for readers with color deficiency: if you look at color bars in Figure 8-1, you will add your deficiency to that already built into those color bars, multiplying the effect unrealistically; for more details, see the note at the end of the caption of Figure 8-1.)

For example, take the top four color bars: they compare normal vision with "red-weak" and "red-blind" vision (with medical names protanomaly and protanopia, respectively). Watch the red colors at left and at right: we see that red-weak people perceive the red as brownish, while red-blind people perceive it as greenish-brown. Given that brown contains red, is greenish-brown what you would expect from strict "red-blindness"? I would expect no red component at all, but that turns out to be wrong! To confirm that, I have used a simple graded filter over the fourth color bar to filter out the reds from the original normal-vision color bar: there you see black (absence of all colors) instead of red components; we can conclude that red-blind does not mean invisibility of red, as we will explain later!

Here is an even greater surprise: look at the green range in the "red-weak and red-blind" group; red-weak people see the greens as yellowish green, while red-blind people see them as orange-yellow! Amazingly,

---

[1] https://en.wikipedia.org/wiki/Color_blindness#Epidemiology.

[2] https://commons.wikimedia.org/wiki/File:Color_blindness.png.

red-color deficiency changes how green is seen! Not only that, but orange and yellow contain red, even though it is being suppressed, not added! Even the blue colors change, although to a lesser extent.

With strict red-blindness, one sees in effect only two basic colors: yellow and blue (and their combinations). Even though yellow can be decomposed into red and green, the red and green are never seen separately.

Similar surprises — and more — await us in the "green-weak and green-blind" group. The green colors are changed into yellowish-green and brownish-orange, respectively, for green-weak and green-blind people (not toward black), even though yellow and orange contain green. And the red colors are again changed toward brown and brownish-green, even though brown contains green.

With strict green-blindness, one again sees in effect only two basic colors, not three: brownish-orange and bluish. So such colorblindness removes one color, although it is not simply the removal of red, green or blue; instead one sees two more complex colors.

The new surprise here is that people with red-deficiency and those with green-deficiency see almost the same colors, whereas a simple-minded color-specific blindness would predict very different perceived colors (compare the color bars labeled "red-filtered only" and "green-filtered only"). This leads to the **red-green color blindness** that we mentioned above: it is indeed common to people with red-deficiency and those with green-deficiency.

We are beginning to see that terms like "red-weak" and "green-blind" are not very accurate or even useful. There appears to be "cross-talk" between colors; more specifically, the eye's cones are not specifically "red-sensitive" or "green-sensitive", but are sensitive to more colors than only "red" or "green". And we encounter red-green color blindness, the difficulty to distinguish red from green.

Continuing to the third group, with "blue-weak and blue-blind" vision, we also find similar surprises. First, blue does not turn into black, but into a greenish blue (dark cyan). And second, the greens turn into light bluish green (cyan)! These effects are however smaller than for the first and second groups: red and green deficiencies appear to be related, as we will discuss in the next Section; also, far fewer people are affected by blue deficiency.

We notice that the blue-weak and blue-blind deficiencies cause far less change in color appearance than the red and green deficiencies. This leads to a useful practical suggestion for **graphical design**: It is recommended that color graphics meant to convey information for general viewing (such as advertisements, brochures and instruction manuals) should minimize using red with green colors; instead they should use primarily red with blue colors, or green with blue colors, since color-deficient people can more easily distinguish those pairs of colors.

Some other deficiencies are much less common. For example, a person could be "double-blind" in two colors, say green-blind as well as blue-blind, leaving visibility only in the red. The "triple-blind" or "all-color-blind" case (called complete achromatopsia) leads to gray-scale vision due to our eyes' rods, while incomplete achromatopsia preserves some limited color vision through a small number of cones.

A major cause of color-weakness is that some of the corresponding cones are inactive; for instance, in green-weak vision, part of the green cones may not actually detect light, so the green cones collectively give a weaker "green" signal to the brain. In the case of red-green color blindness, another cause is also common: the red and green cones are more similar to each other (specifically, their peak sensitivities occur very near the same color), so they send similar signals to the brain, which then cannot easily distinguish red from green.

In Section 8.2, we will discuss in more detail how all this works. That will also teach us interesting and useful aspects of normal vision.

8.2   Surprising lessons for normal color vision

In Section 8.1, we came to strange conclusions: for example, red-blindness changes red colors to browns and green colors to orange-yellows, even though brown, orange and yellow all contain red components. Also, people with red-blindness and people with green-blindness experience very similar red-green color blindness, namely difficulty to distinguish red from green, even though you would think that removing red *versus* green should give very different results. ***How could all that be true?***

To understand this, we have to go back to our eyes' cones. Throughout this book, we have assumed that, in **normal vision**, "red cones" are sensitive to red light, "green cones" to green light and "blue cones" to blue light. We now discover that reality is more complex: in fact, "red cones" are normally sensitive to all red, green and blue light; and so are green and blue cones. Therefore, the difference between the three cone types should be stated more precisely as: "Red cones" are predominantly sensitive to red light, "green cones" predominantly to green light, and "blue cones" predominantly to blue light, while each cone is also sensitive to the two other basic colors.

Another way of saying this is that the sensitivities of the red, green and blue cones overlap. This means, for example: if we shine red light into our eyes, all cones will detect that red light — not only the red cones, but also the green cones and the blue cones.

There is especially strong overlap between red and green (much more overlap than between blue and either red or green), as detailed below. It is then not too surprising that a minor deficiency in red or green cones can in fact produce red-green color confusion.

There is another amazing consequence of the color overlap between cones. As an example, take pure red light shining onto the cones. As we have seen, pure red will not only excite the "red" cones, but also the "green" and "blue" cones. Normally, the red cones will be excited most intensely by red light, then green somewhat less and blue even less: the relative intensities of excitation are the signals sent to the brain to represent pure red. So we have the paradoxical situation that pure red (such as the laser light of Figure 2-10) causes not only red, but also green and blue signals to be sent to the brain; the brain in turn must then translate this mix of signals to tell you that you are actually looking at simple "pure red" light, with absolutely no contribution from green or blue! This seems not only paradoxical, but also highly inefficient; and yet, that is how our vision operates!

As a result, for simple "pure red" light, the cones send complex signals that are mixtures of red, green and blue intensities, not a simple signal that says this light is "pure red"; and yet, the brain translates these mixed signals into the message "you see pure red".

The same holds for pure green light: it causes another set of red, green and blue signals to be sent to the brain, although now with different relative intensities (more green but less red and blue), which the brain translates into "pure green". Pure blue light will also cause a set of red, green and blue signals, but with more intensity in the blue and less in green and red, which the brain translates into "pure blue".

The situation becomes even more complex with incoming light of mixed color content, for example yellow, which contains both red and green: this mixes together the signals expected for pure red and pure green. The brain now needs to disentangle those even more complex signals to tell you that the incoming light was yellow.

Thus, such overlaps or "cross-talk" in color sensitivities between the cones have to be unraveled by the brain: the brain must somehow assign the correct color to the detected light, even when that light is purely red, green or blue, despite the fact that it excites cones of more than one color. This is not a simple task, but happens constantly in normal color vision as well as in color-deficient vision. An important consequence for color blindness is the following: even if, for example, the green cones don't function at all (they are strictly "green-blind"), green light can still be detected by our red and blue cones.

The above paragraphs are probably very confusing. **What is going on?** We reached a contradiction between the colors that the eyes detect and the colors that the brain sees: the colors that we see in our "minds" are not those that the eyes see, based on simply comparing the signals coming from the eyes' cones! **Can you guess why?**

We have already several times hinted at a most important factor in vision: the **brain** itself, or more exactly the entire nervous system that leads from the cones in the retina to and through the different parts of the brain that convert the detected signals into an image in your mind. Indeed, the brain performs many functions (mostly without our knowledge) before delivering the finished image for you to view on a virtual screen somewhere inside your mind! It is like a modern digital camera that processes the detected image in many ways before presenting it for your viewing, except that the brain does even more processing than a camera.

READ HERE FOR A MORE DETAILED UNDERSTANDING OF HOW COLOR PER-CEPTION WORKS: We now need to ask: **How sensitive is each cone to each color or hue?** This is shown schematically in Figure 8-2 for the solar spectrum (the solar spectrum is the major part of the color bars used in Figure 8-1). The red curve shows the **sensitivity** of the "red cones" to light of different colors (or hues, as quantified by the wavelength of light marked along the bottom). Similarly, the green curve shows the sensitivity of the "green cones", and the blue curve corresponds to the "blue cones". For comparison, the sensitivity of the rods is also shown as a dashed gray curve.

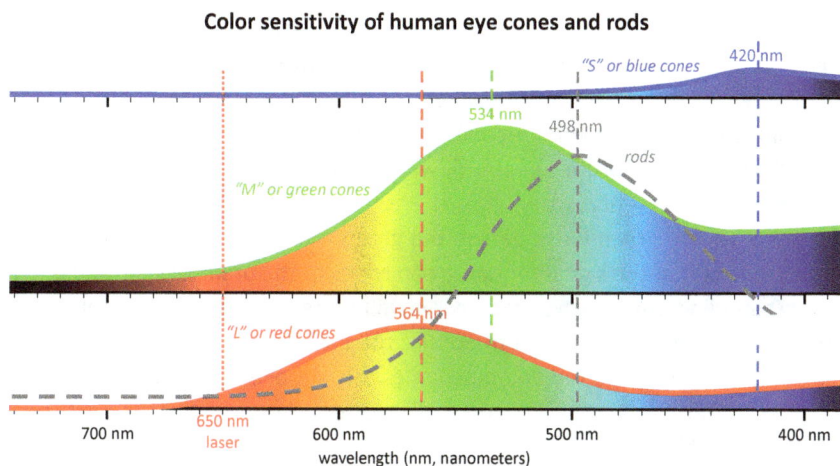

Color sensitivity of human eye cones and rods

Figure 8-2: The three red, green and blue curves show the approximate normal sensitivity of the red, green and blue cones to the solar spectrum in humans, while the gray curve represents the rods (the spectral colors below the curves are the same as shown in Figure 2-11). A peak in a curve indicates maximum sensitivity of the corresponding cones to light of that color (hue or wavelength): for example, the green peak labeled "534 nm" indicates that green cones are most sensitive to light which has a wavelength near 534 nanometer; this lies in the middle of the green part of the solar spectrum. However, the red peak is clearly <u>not</u> in the middle of the red part of the solar spectrum, but closer to the green part! The "wings" of the peaks to their left and right indicate that the cones retain sensitivity to light even if its hue (wavelength) is not close to the peak. In particular, the green curve extends all across the visible range. Because of these extended ranges, the red, green and blue cones are often labeled "L", "M" and "S" for long, medium and short wavelengths, respectively. The heights of the peaks indicate the relative sensitivities of the red *versus* green *versus* blue cones, given by the weights 0.3:0.6:0.1 (more precisely 0.299:0.587:0.114 on average); this shows that our green cones normally have about twice the sensitivity of our red cones, and about six times the sensitivity of our blue cones. (The rod sensitivity is not drawn to scale.) Many such curves exist on the web, with great variability: the curves shown here are only schematic, not universal; they are redrawn from https://commons.wikimedia.org/wiki/File:1416_Color_Sensitivity.jpg under the Creative Commons Attribution 3.0 Unported license.

(Continued)

(Continued)

We here continue to use the names "red cones", *etc.*, because they are convenient: however, as concluded above, we must remember that these names mean "predominantly sensitive to red", *etc*. There exist more correct names, but they are less evocative and less easy to remember: as labeled in Figure 8-2, the "red cones" are called **L cones** by scientists, because these cones are more sensitive to "long" wavelengths, meaning to the red end of the solar spectrum; the "green cones" are called **M cones** to indicate their greater sensitivity to the "middle" of the spectrum, where green is concentrated; and the "blue cones" are named **S cones** because they are more sensitive to the "short" wavelengths of the spectrum, where blue is found.

The sensitivity curves of Figure 8-2 relate as follows to color-weakness: for example, in a green-weak person the green curve will be depressed, which simply means that the signal sent by green cones to the brain will be weakened. In a green-blind person, the green curve will be reduced to zero, because the green cones send no signals to the brain: the green cones are then effectively blind. Nevertheless, the red and blue cones can still pick up some green signal.

There also exists another source of color deficiency, discussed next: **sensitivity peak shifts**.

It is striking in Figure 8-2 how much the red and green peaks overlap: basically, the red peak is misplaced toward the green region of the solar spectrum; the red sensitivity curve actually peaks well inside the green range instead of in the red range. As we mentioned in Section 5.1.4, early humans only had two kinds of cones in their eyes: "red" and blue. The "red" cones later split into today's red and green cones, with relatively much overlap. This can explain one common source of color vision deficiency: the red and green peaks in some humans are shifted so close together that it is difficult for them to distinguish red from green. The result is similar to the red-green blindness mentioned above, but in that case due to weakness of red or green cone response.

All the above-mentioned observations indicate that our brain interprets the signals from our cones in a remarkably complex manner.

8.3 Variations in normal color vision

We have discussed above how color deficiency differs from **normal vision**. That has also taught us about normal vision itself. But we may also ask: *How variable is normal color vision?* After all, each person is different: just as our bodies have different shapes and abilities, our eyes also vary. In fact, the eyes' abilities vary not only between different people but also within a single person. This aspect will lead us to some new surprises!

What variations exist between people? As we know, color vision initially comes from the color-sensitive cones in the retina of our eyes. Our cones naturally vary from person to person.

However, the way our **brain** treats the information it receives from our eyes also affects our perception of color. And the brain's perception of color is also affected by what you see as you grow up after birth, because the brain learns from experience (examples will be given below). This learning thus depends on your experiences in viewing the outside world (such as whether you grow up in a concrete-and-bricks inner city or a green wooded neighborhood). Learning is also influenced by what society tells you about colors and color names; in particular, color-deficient people are in the minority and thus must learn to conform with the majority's notions of colors and color names. Learning therefore creates many opportunities for large variations between people: vision is thus partly subjective.

How about variations within a single person? Of course, there may be differences between our two eyes, and changes over time. More interestingly, consider a simple example: light intensity. As you know, the **pupil** (the opening that lets light enter the eye, see Section 2.4) can adjust its size so that the intensity falling into the eye can vary about fourfold between bright and dark sceneries: this is an automatic effect that you cannot control. Thus, the same object can look the same to you even if the light intensity changes fourfold. In particular, it will keep the same apparent color: your brain is indeed able to compensate for very large variations of light intensity.

A more subtle example: the color white. It is remarkable that we can agree so well on what the color of a white page is: that is actually far from obvious! As we know (Sections 2.5 and 4.1), white light can be seen as a mix of equal amounts of red, green and blue light. But, if the cones in your eyes do not see the mix as quite equal, the "white" will contain a bit too much — or too little — red, green or blue color. Then white looks **off-white**, such as pinkish or greenish white, so that every page that you see would have an off-white coloring. Figure 8-3 shows some off-white colors, based on a 10% excess or deficit of red, green

```
white: (R,G,B) = (255,255,255)




off-whites (+10%):
reddish (255,230,230)        greenish (230,255,230)        bluish (230,230,255)




off-whites (-10%):
light cyan (230,255,255)     light magenta (255,230,255)   light yellow (255,255,230)
```

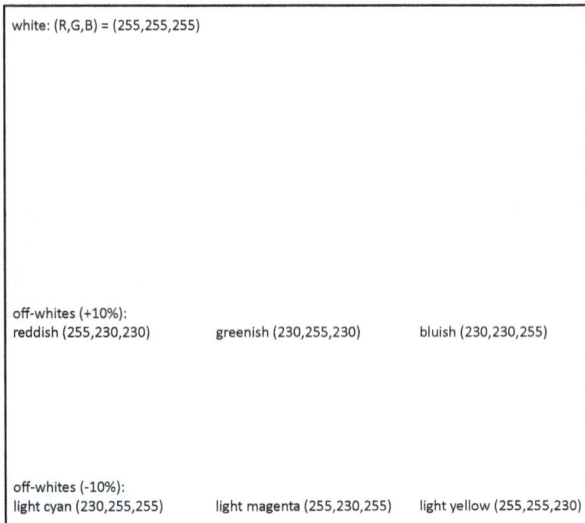

Figure 8-3: Comparison of "pure white" with off-white, obtained by 10% excess or deficit of red, green or blue (labeled with RGB components). Try to imagine an entire room painted in an off-white color: it may look white to you.

or blue. This effect should be quite common as our cones are rarely perfect: very likely our cones have slightly different relative sensitivities to red, green and blue; also, as we saw in Section 5.1.3, cataract surgery that replaces the eye's lens can change the color of light, requiring the brain to compensate. Nevertheless, we surely agree that the "white" paper used in books, newspapers, brochures, *etc.* is properly white, not off-white.

Another example is the color of "white" walls in many buildings: very few walls are painted with "proper white" (because that white is too harsh). Instead, off-white paint is often used, but we still view it as white (compare wall colors with white paper!).

These examples illustrate that our brains automatically compensate for deviations from white and make us believe that we are seeing proper white. This must be stressed: our vision cannot be perfect, so our brains automatically compensate to make off-white look white.

We can extend this example of white paper: the lighting source under which we see the paper can also vary in its coloring, giving a non-white reflected color (see Section 4.2). For example, reading lamps tend to give a relaxing yellowish light, while most fluorescent and LED lamps produce a harsher, slightly bluish light. Nevertheless we will still claim that the paper is white, instead of concluding that the paper color has magically changed from slightly yellow to slightly blue in going from one room to another. Here again, the brain is automatically compensating by adjusting the perceived color to an assumed universal pure "white" color. This compensation is what digital cameras also try to do automatically (although not always correctly, see Figure 2-2) when they perform white balancing.

The above examples show that "normal" vision is indeed quite variable, both between people and within one person. But we next address a more subtle and vexing question, which is particularly relevant when comparing what colors different people see and when communicating about colors with color-deficient people.

How do you describe in words the colors red versus green versus blue, as shown for example in Figure 2-13? One approach is to say that red is the color of ripe tomatoes, green the color of fresh plants, and blue the color of the clear sky; that is true but very vague (do tomatoes, plants and skies have the same color everywhere?). Another approach is to give the wavelength of the corresponding electromagnetic radiation; this is also true and much more precise, but not practical (how do you measure the wavelength of light and produce light of that wavelength?).

However, those answers avoid the harder questions: *How can we say in words how different red is from green or blue? Does red look the same to you and me? And how do we tell each other what red looks like? In particular, how do we tell a green-blind person (whose "green" cones do not respond) what green looks like? Similarly, how do we tell a totally blind person what any color looks like?* I have not been able to find words to describe what red looks like to me, and likewise for green or blue. Yet I am not color-blind, and I do have a very vivid vision in my mind of the colors red, green and blue. This leads to a surprising conclusion, even if you and I have normal color vision: the same color that I see as "red", you may see as any other color, such as my "purple"; and what you see as "green", I may see as your "cyan", for example. Another random example: I may see the complementary colors (see Section 2.7) of the colors you see! Why not? In other words, we have no certainty that you and I perceive the same color in the same way:

the colors you and I see could well be scrambled or even quite different! I am not aware of any test that can "measure" how we see a particular color. Perhaps, color perception is very subjective.

This also has profound implications for communicating between people with colorblindness and those with normal vision. How can color-blind people tell a person with normal vision what colors they see? And how would a person with normal vision tell green-blind people what green looks like? We have also discussed these issues in Section 7.6.

This difficulty of communicating about colors makes it hard to study and discuss color deficiency.

8.4 What have we learned in this Chapter?

Very few people are actually "color-blind": most of them see one color relatively weakly. The terms "color vision deficiency" or "color deficiency" are therefore preferred.

We found that the "red", "green" and "blue" cones are normally all sensitive to all colors, not just to one component of light: for example, "red" cones also detect pure green and pure blue. This overlap or crosstalk is largely responsible for red-green blindness, in which red and green are difficult to distinguish.

On the other hand, the brain plays a complex role in interpreting the signals it gets from the cones, whether in normal vision or in color-deficient vision. Even "pure red" in our mind is the result of disentangling by our brain of mixed signals from the "red", "green" and "blue" cones. The brain also manipulates colors to compensate for variations in illumination: a white page in a book will look to us white under a wide range of circumstances.

A challenge is telling other people (whether with normal or color-deficient vision) what a given color looks like: it may well be that "pure red" actually looks different to different people! And it is impossible to tell a truly color-blind person what that missing color looks like.

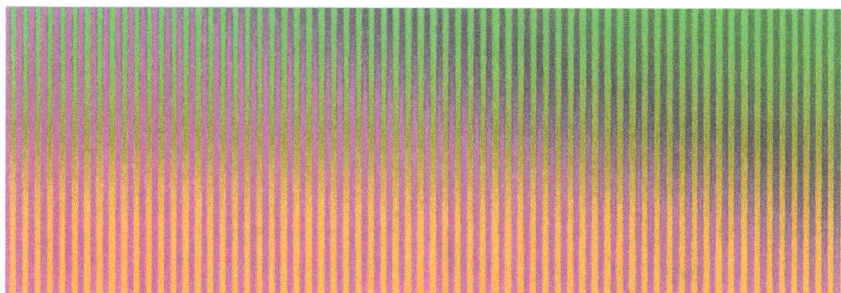

9

Wavy Artifacts: Moiré Patterns

*You surely have seen fast-moving wavy patterns in sheer curtains or on some computer displays, but you probably did not worry much about them, as they are neither harmful nor useful. Nevertheless, they can be quite distracting and intriguing. Such patterns are actually fairly common: in this Chapter we will show and explain a variety of interesting examples. They are often entertaining, but there are some disturbing cases as well, such as horrible pictures of patterned clothes! We will use animations[1] to show the origin and character of the rapid movements in these patterns. A video is also available online to illustrate the most important points.[2] These **wavy artifacts** are called **moiré patterns**, where the French word moiré means wavy.*

⇐ ·}} {{· ⇒

---

[1] Animations are available to buyers of this book at https://worldscientific.com/world-scibooks/10.1142/12316#t=suppl (for more details, see References and Resources on page 257).

[2] See video "Wild wavy webs: moiré patterns" on Everyday Physics by Michel A. Van Hove at YouTube: https://youtu.be/QPNNzBtKFzo.

9.1 Examples of artificial wavy patterns

Most likely, you have seen funny wavy patterns in **sheer curtains** (semi-transparent draperies), like those at left in Figure 9-1. You may have wondered: ***What are these wavy patterns due to and why do they "move" so fast all the time?***

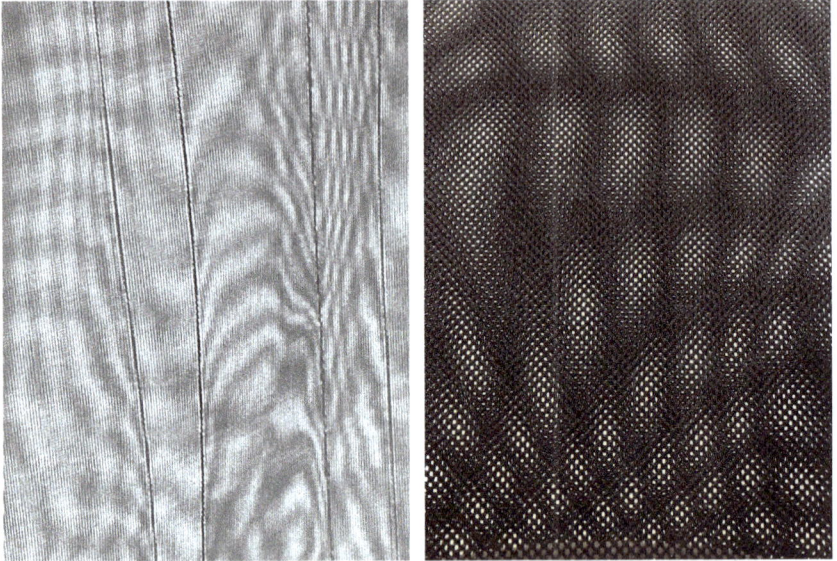

Figure 9-1: The sheer curtain photographed at left shows a wavy pattern called moiré pattern; the photograph was taken through three layers of curtain (it is rare to look through just two layers of the same curtain!); the dark lines are seams in the curtains which are not part of the moiré pattern but which break up the wavy pattern into independent regions.

The chair back photographed at right also shows a wavy moiré pattern; it is formed by two almost identical semi-transparent black nets spaced by about 1 centimeter; at bottom, a narrow strip shows only one of the two nets, without the wavy moiré pattern.

If you stretch out a sheer curtain so that it is a single flat layer, do you see such a pattern? Most likely not: you just see a regular square mesh of threads, but no hint of a wave. But if you fold a sheer curtain so you look through it twice (or more times), suddenly the wavy pattern appears! Thus, the pattern is a combination of the meshes of threads in two (or more) layers of sheer curtain: the pattern is therefore an artifact that was not present in the curtain itself.

The wavy pattern actually "moves" as you move your eye, or if you move one curtain layer relative to the other. This also shows that the pattern is an artifact.

Also very noticeable is how fast these patterns move as you slightly move your eye or the curtain layers. A slight breeze or your breath can make those patterns race around the curtains!

If you look closely, you will find that the wavy pattern is due to the alignment *versus* misalignment of threads from the folding of the curtain into different layers. **The pattern is therefore an "artifact" or "visual combination" of two (or more) regular meshes.** Such a pattern is often called a **moiré pattern**, from the French word moiré that means wavy (moiré is pronounced as "mwa-ray").

At right in Figure 9-1, I show the back of my office chair. It is made up of two layers of netting with a space between them, similar to layers of sheer curtains. This chair back gives a very strong moiré pattern, because the two nets can totally block the view through them when they are suitably aligned, while remaining partly transparent with other alignments.

Some aspects of these wavy moiré patterns are already obvious; we will see them in other situations as well. They include: **The moiré waves are smooth and gentle; the waves may be straight or curved; there may be waves in more than one orientation; the wave orientations are not simply related to the mesh orientations of the individual layers; the waves have clear wavelengths** (the wavelength is the distance between similar contrasting regions, like dark regions or bright regions); **the wavelength may vary across the pattern and may vary with direction; the wavelengths are much larger than the meshes of the individual layers**.

The wavy moiré patterns shown in Figure 9-1 are fun and intriguing to look at. **Sometimes, however, moiré patterns are very unpleasant.** See the photograph of a suit at left in Figure 9-2: it has a finely detailed mesh in the cloth, better seen in the blow-up below. At right is an ugly photograph of the same suit, simply taken from another distance, where the fine mesh of the cloth clashes with the regular mesh of recording pixels in the digital camera. If you want to advertise patterned clothing, you should definitely avoid this disastrous moiré pattern; fashion photographers know this very well!

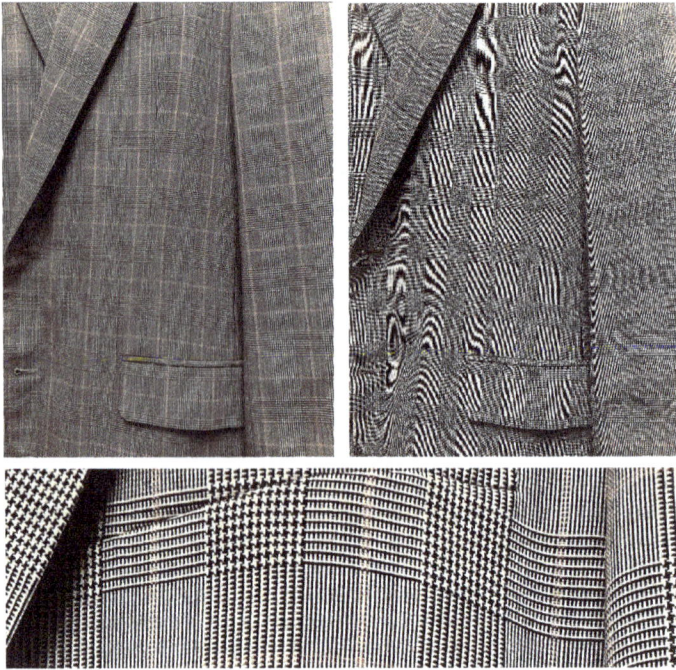

Figure 9-2: Digital photographs of a patterned suit (top left), showing an ugly moiré pattern (top right), and detail (below).

You can easily produce this ugly effect yourself: you may change the size of Figure 9-2 if you view it on a display, or you may view a finely patterned piece of clothing through a digital camera while you slowly move the camera forward and backward. You will find that such wavy moiré patterns appear in a certain range of distances. You will also notice that the moiré pattern is highly variable as you move the camera! Thus, it is easy to avoid such moiré patterns: they are highly visible, and you only need to move the camera closer or farther.

As mentioned, the moiré patterns move and change very fast as you move the eye or camera even a little bit: such high sensitivity and mobility is found with all moiré patterns, as we will explain in Section 9.2. In fact, you may see some flashing of the images in Figure 9-1 and Figure 9-2, as you scroll the page or change its magnification: this is also due to the high sensitivity of moiré patterns, in this case because of the combination with the mesh of display pixels of your computer display.

Again, the wavy moiré pattern is due to the visual combination of two (or more) independent meshes: the moiré pattern does not exist in the cloth alone, or in the camera or display alone. The moiré pattern is an artifact which only appears when you view the two (or more) independent meshes together.

Figure 9-3: Digital photograph of a patterned pair of apartment buildings, showing a variable moiré pattern as the pixel size is increased from top left to lower right. The pixel numbers drop from 527 × 549 at upper left to 141 × 154 at upper right and 58 × 68 at lower right.

The same effect is shown in Figure 9-3, where I photographed two distant apartment buildings and varied the pixel size of the final photograph. The top left version of the photograph looks good, even when magnified, because it has small pixels. But the other versions with larger pixels show increasingly disruptive moiré patterns, ending in disaster when each pixel is close to the apparent size of an apartment. Once again, we see the artificial combination of two fine meshes: here the regular mesh of apartments and the regular mesh of image pixels.

Another dramatic example is shown in Figure 9-4. Here a photograph of a simple drawing (the original is shown at left) is displayed on a

Figure 9-4: A simple drawing (at left) on a computer display, at two slightly different magnifications (at center and right): the original drawing (left) is entirely composed of uniform patches of orange, white and black (the camera automatically balanced the colors, turning white into gray).

computer screen in two versions (at center and right): by selecting a suitable magnification of the image on the screen, an intense wavy moiré pattern is generated (center); and a slight change of the magnification completely changes that pattern (right). It is remarkable that the original drawing shown at left consisted only of uniformly colored shapes, with no wavy pattern at all: the artificial moiré pattern is entirely due to the visual combination of a mesh of image pixels and a mesh of display pixels. Note that uniform black produces no wavy pattern, because black means that no light is available to make any pattern, whether regular or wavy.

We can also easily observe wavy moiré patterns by viewing an electronic display with a digital camera. Figure 9-5 gives two pretty examples. Here I have simply photographed a smartphone display and a television screen, both showing only a white page.

You can do this yourself (you may look at any patch of uniform color, not only white): move your camera in front of such displays until the moiré pattern appears; you will have to search for the right distance range, which depends on the exact dimensions of your smartphone pixels and your screen display pixels. Again, the moiré patterns will move very fast as you move the camera even a little bit, so be patient and keep your camera steady (perhaps on a tripod or support).

To avoid unpleasant moiré patterns, most digital cameras and displays have software (for example in the graphics card) that tries to eliminate them. Without such built-in software, we would see many more artificial and disturbing wavy patterns on our computer displays, smartphones and television screens.

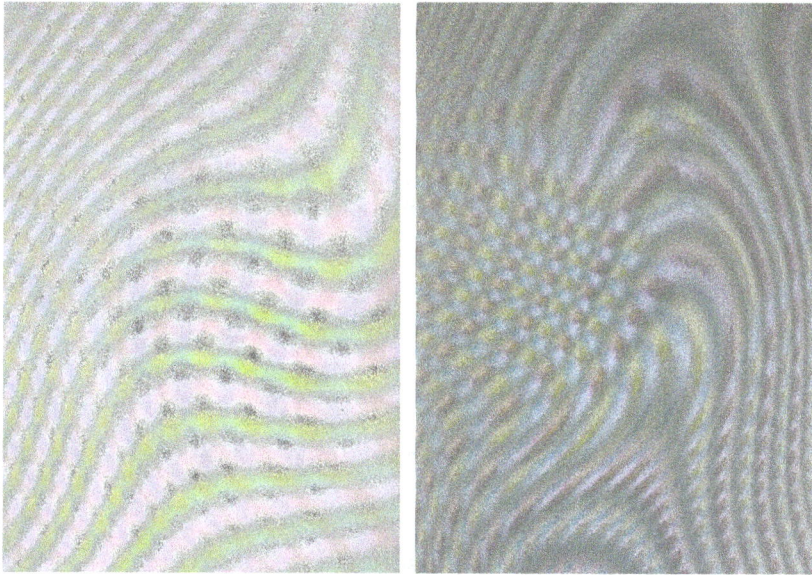

Figure 9-5: Digital photographs of the white screens of a smartphone and a television set, at left and right, respectively (the camera automatically balanced the colors, turning white into gray).

9.2   One-dimensional moiré patterns

How does a wavy moiré pattern arise? It will be helpful to first look at simple **moiré patterns in one dimension**, namely in one direction only: they are already very interesting and informative. In Section 9.3 we will then discuss patterns in two dimensions, like the images we saw in Section 9.1.

You can use a simple comb in front of a mirror (or in front of its sharp shadow on a wall or page) to create a one-dimensional moiré pattern, as shown in Figure 9-6. The comb and its mirror image (or shadow) create an artificial repeating wavy pattern in one direction: it is a **moiré pattern**.

It is clear in Figure 9-6 that the pattern is composed of lighter and darker bands. The lighter bands arise when the teeth of the comb and its image overlap and thus allow the light to shine through the gaps between the teeth. By contrast, in the darker bands the teeth of the comb block the gaps between the teeth of the image comb (and *vice versa*), so no light can pass through.

Figure 9-6: We see here a comb and its image in a parallel mirror producing moiré patterns. The top two photographs differ in the distance between the comb and the mirror, and therefore the distance between the comb and its image. In the bottom photograph the comb is tilted such that its top is closer to the mirror and to the mirror image.

The pattern changes when the distance varies between the comb and the mirror (and therefore also the distance between the comb and its mirror image), as seen in the two upper photographs: the top photograph has 9 dark bands, while the middle photograph has 8 dark bands. The reason is that the image comb becomes relatively smaller than the comb itself as it becomes more distant, so that overlaps occur sooner, as we will discuss below. The lower photograph combines the two cases by tilting the comb: the comb's upper part is closer to the mirror and to its image than the lower part, so that the number of dark bands varies; this produces inclined dark bands.

We now illustrate the example of the comb more precisely with drawings of two perfect fences, as shown in Figure 9-7. (You can imagine the gray fence to be the mirror image of the black one, similar to the comb in Figure 9-6, or you may imagine looking at two separate parallel fences, or at a fence and its shadow on a parallel wall.)

In Figure 9-7, the gray fence is farther back, so it looks a bit smaller. Where the fences overlap, we see the same artificial dark bands that we saw with a comb and its mirror image: we again have a moiré pattern of darker and lighter bands.

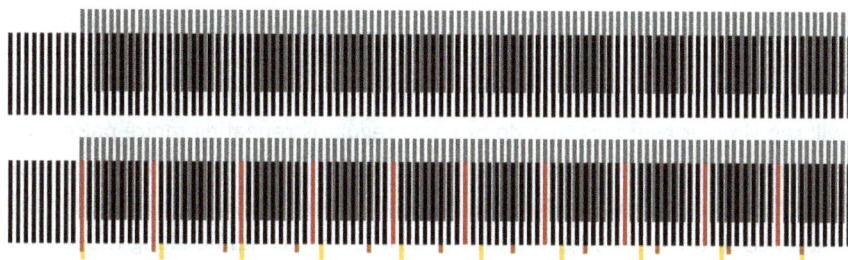

Figure 9-7: Top image: a black fence stands in front of a more distant gray fence; the gray fence therefore looks smaller. A wavy moiré pattern is formed where the two fences visually overlap. Bottom image: the same two fences, but here the black fence slats that closely overlap gray slats are colored red; also, the brown markers mark every 9th slat in the black fence, while the yellow markers mark every 10th slat in the black fence: these markers allow us to see irregularities in the widths of the light and dark bands.

As with the comb above, these two fences are made of slats ("teeth") that have about the same width as the gaps between them. As a result, when black and gray slats overlap, you can see through the gaps in both fences. But when the slats do not overlap, they block the view through both fences because each gap of one fence is covered by a slat of the other fence. This effect gives the contrast between the bright and dark parts of the moiré pattern. And this is also the basic principle behind all the examples we saw in Section 9.1: Regions of overlap contrast with regions of non-overlap to form the wavy moiré patterns.

The lower part of Figure 9-7 repeats the same pair of fences, but points out where overlaps occur: the red-colored slats of the black fence are those that most closely overlap a slat in the gray fence. The dark parts of the moiré patterns (where slats do not overlap) fit neatly between those red slats.

HERE IS A SLIGHTLY MORE TECHNICAL QUESTION: *Are the red overlapping slats in Figure 9-7 regularly spaced?* At first sight, the red slats do appear to have equal spacings between them. However, if we count those spacings precisely, we find irregularities! (The regularly spaced brown and yellow markers at the bottom of Figure 9-7 allow you to easily count slats: they show quite clearly that the red slats are not regularly spaced.) Indeed, the red slats are spaced by 9, 10 or 11 slats, depending on where you count them. The average spacing is about 9.67. Where does this strange average come from? I purposely chose that number when constructing the two fences! I did not want a simple average like 9 or 10, but something more random, like you might experience when looking at a real pair of fences or a comb in a mirror. In Figure 9-8 and Figure 9-9, you will see simpler averages that do produce regularly repeating moiré patterns.

Irregular spacings in a moiré pattern are quite normal: you should in general expect such irregularities. Moreover, normally there is also no repetition in the sequence of these spacings: no matter how far you continue along the fences, the sequence of spacings will usually not repeat itself (in Figure 9-7 the spacings between red slats are 9, 11, 9, 10, 9, 10, 10, 10, 11, *etc.*). Scientists call these situations aperiodic or incommensurate lattices, with length ratios that are irrational numbers (namely, numbers that cannot be written as the ratio of two whole numbers). In "lucky" cases, there is repetition of the spacings, as we will see next. Scientists then use the terms periodic or commensurate lattices, and rational numbers that can be expressed as ratios of whole numbers.

In Figure 9-7, the spacings between the red overlapping slats were somewhat irregular (as discussed in the more technical note above). *Can the spacings between overlapping slats ever be regular, namely repetitive?* The answer is yes, as shown in both Figure 9-8 and Figure 9-9. In these two figures, the gray fence has two different sizes relative to the black fence. In Figure 9-8, 9 black slats fit exactly 10 gray slats, while in Figure 9-9, 19 black slats fit exactly 20 gray slats: overlaps are shown by red and blue slats; the overlaps continue to the right forever with the same regularity. These are thus repetitive, or "periodic", moiré patterns. For these examples, the ratios of slats are 9/10 and 19/20 (such ratios of whole numbers are called rational numbers).

Where does the high sensitivity and mobility of moiré patterns come from? We have mentioned several times in Section 9.1 how

Figure 9-8: A black fence stands in front of a more distant gray fence, such that every 9th black slat exactly overlaps every 10th gray slat. The red and blue slats show some of the overlaps, which continue with perfect regularity forever to the right.

Figure 9-9: As Figure 9-8, but here every 19th black slat exactly overlaps every 20th gray slat.

sensitive moiré patterns are to small movements of your eye or of the meshes that you are watching. Figure 9-10 illustrates how this happens.

For better visibility, I have here drawn a lower gray fence in front of a higher black fence and repeated the gray fence with very small displacements to the right: the blue slat of the gray fence moves repeatedly by a quarter of the distance to the next overlap with a black slat. As that motion happens, the moiré pattern moves much farther to the right. In this example, as the blue slat moves to the next black slat, the orange overlaps race forward by 19 black slats, meaning with a 19 times larger speed. The entire moiré pattern moves to the right at that much amplified speed.

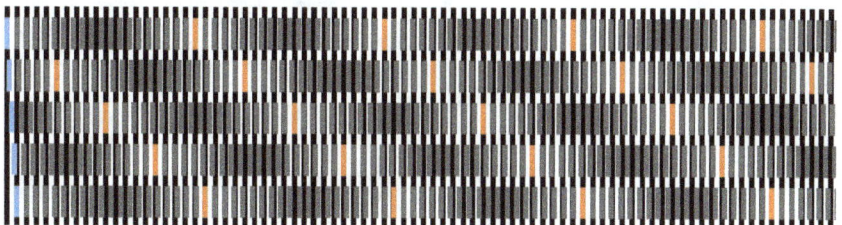

Figure 9-10: Effect of a slight movement of the gray fence, shown as five shifted thin gray strips in front of the black fence. As in Figure 9-9, 20 gray slats fit every 19 black slats. From top to bottom, the blue slat of the gray fence is moved four times by a quarter of the distance until it overlaps the next black slat to its right. The orange-colored slats mark the overlaps of gray and black slats, showing a 19-fold speed increase of the movement of the moiré pattern relative to the speed of the fence itself.

This same effect is used in devices called **micrometers** (or **micro-meter screw gauges**) to obtain high precision in measuring lengths. The principle is simple: the large motion of the orange slats in Figure 9-10 amplifies 19-fold the tiny motion of the blue slat, so the position of the orange slat gives a 19 times more sensitive measure of the blue slat's position. In micrometer screw gauges, a small length difference converts into a large rotation of a screw, resulting in a large change along a gauge, which is more easily measured.

This effect is exhibited in Animation 9-A1, which you may view by downloading a set of PowerPoint files (*see* footnote 1 in this Chapter). In that Animation 9-A1, a screenshot of which is shown next, the same black fence is moved back and forth across the gray fence: while the black fence moves by only one slat, the wavy moiré pattern moves much farther.

This speed amplification is a quite general result for moiré patterns: The size of the moiré pattern (namely the wavelength as measured by the distance between equal overlaps, for instance), divided by the size of the individual mesh unit, gives the speed multiplication factor, here 19/1 = 19.

Animation 9-A1 screenshot: The animations are available for download (*see* footnote 1 in this Chapter).

You may have noticed a surprising contrast between the Anima-tion 9-A1 and Figure 9-10: the initial direction of movement of the moiré pattern is opposite, namely to the left in Animation 9-A1 and to the right in Figure 9-10. This contrasting behavior can be traced to the fact that in Animation 9-A1 the moving mesh (black) is larger than the (gray) stationary mesh, while the opposite is true in Figure 9-10: there the smaller mesh moves. This is clearly visible in Animation 9-A2, which combines both cases. It also compares them with the situation where the front and back fences have the same slat spacing: then no moiré pattern occurs; there is only a smooth blinking between bright and dark.

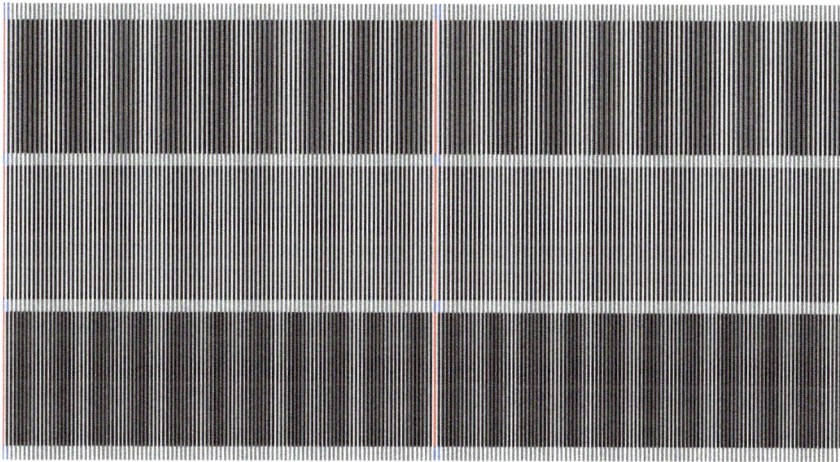

Animation 9-A2 screenshot: The animation is available for download (*see* footnote 1 in this Chapter).

We see that **the direction of motion of the moiré pattern can vary from case to case**. More surprises will be found in the two-dimensional case: see Section 9.3.

What happens when we add a third layer, like a third fence or another fold in a sheer curtain? Remember that when you look through sheer curtains, you are usually looking through three layers, not two, as we have discussed so far (the reason is that one fold must be followed by another reverse fold for the curtain to continue in the same direction). So let's add a third fence to the two fences of Figure 9-7: in Figure 9-11, I have added a blue fence in front of the black and gray fences; being

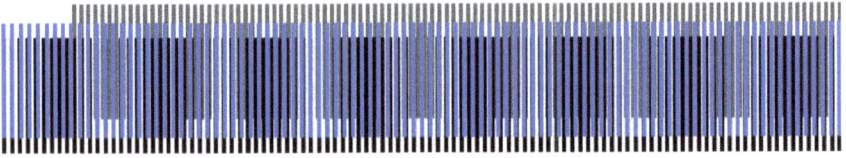

Figure 9-11: Three fences: blue in front, black in the middle, gray behind. The three fences are shifted vertically to better show their individual mesh structures.

in front, this blue fence looks stretched out a bit compared to the black and gray fences.

What kind of moiré pattern do we get now? The central horizontal strip of Figure 9-11 shows the superposition of the three fences: the result is much more irregular than in Figure 9-7.

You can barely recognize here the former moiré pattern of Figure 9-7, as it is mixed into two new moiré patterns: the moiré pattern due to the combination of blue and black fences and the pattern due to the blue and gray fences. The result is therefore much wilder. Moreover, this triple combination of pairwise moiré patterns is highly sensitive to any changes. You may view an animation of this effect in Animation 9-A3.

That high sensitivity is why you see moiré patterns in sheer curtains dance around so much in the slightest breeze or breath. An additional reason is that curtains are curved, not flat like our simple examples here, so the patterns additionally change with the curvature of the curtains.

Interestingly, the complexity of 3-mesh moiré patterns can be used to combat counterfeiting (fraud). Indeed, whenever one copies, scans or photographs a 2-mesh moiré pattern, a third mesh is added, namely the detector mesh of the copier, scanner or camera (since the pattern is thereby cut up into pixels). That creates a new 3-mesh moiré pattern that is easily observed. The basic idea then is to include in an object that should not be copied (think banknotes!) a 2-mesh moiré pattern. To circumvent this anti-fraud feature would require computational analysis and reconstruction of the original 2-mesh moiré pattern. (This idea was proposed in a patent.[3])

---

[3]US Patent No. 9,633,195 filed on April 25, 2017, https://patents.justia.com/patent/9633195.

Animation 9-A3 screenshot: The animation is available for download (see footnote 1 in this Chapter).

9.3 Two-dimensional moiré patterns

Having discussed the behaviors of one-dimensional wavy moiré patterns in Section 9.2, we can move on to **two-dimensional moiré patterns**, like those illustrated in Section 9.1.

What is new with 2D moiré patterns? Two-dimensional moiré patterns include the behaviors we have just discussed in Section 9.1. However, they add intriguing new behaviors: we will concentrate on those interesting new features in this Section.

First, we show at left in Figure 9-12 a simple 2D version of the 1D fences of Figure 9-7: we recognize the wavy dark and light bands of a moiré pattern in both the horizontal and vertical directions. This suggests, as we will confirm later, that the 2D moiré pattern is simply the combination of the two 1D moiré patterns.

This example uses a black square mesh, whereas I shrank the blue mesh differently in both directions: the blue mesh is thus rectangular, not square, so that the moiré pattern also has a rectangular appearance.

At right in Figure 9-12 we see a similar pattern: in fact, the only difference with the left pattern is that the lines have been replaced by dots, one dot at each intersection of lines.

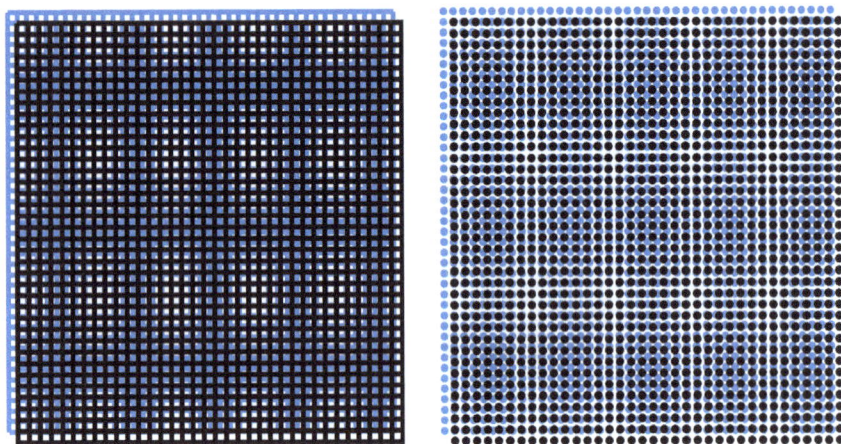

Figure 9-12: At left are vertical and horizontal fences, making up a simple square mesh (black) and a rectangular mesh (blue). The blue fences are more narrowly spaced than the black fences. At right, the meshes of lines are replaced by identical meshes of dots.

Do you recognize similarities between the left and right patterns of *Figure 9-12***?** Three main aspects of the moiré pattern remain identical whether we use lines or dots: the **wave** shape of the pattern is the same (rectangular in this case), the wave orientations are the same (parallel to the lines and rows of dots), and the wavelengths of the pattern are the same (the horizontal and the vertical repeat distances in this case). So, we learn that the waves of the moiré pattern do not depend on the detailed contents of the meshes, such as whether they are lines or dots: The 2D moiré pattern has the same wave shape and the same sizes (wavelengths) regardless of the contents of the 2D meshes.

We see in Figure 9-12 that the version with dots gives better insight into some details of the moiré pattern. For that reason, we will sometimes use dots in the following illustrations.

What happens if we let one mesh smoothly change its size? In Animation 9-A4, illustrated next, we can see how strongly the moiré pattern changes when one mesh (blue or red) grows and shrinks. This shows that the pattern size (wavelength) is very sensitive to the size difference between the two meshes. Notice how the wavelength of the moiré pattern is large when the two meshes have similar size; the wavelength shrinks rapidly when the two mesh sizes become more different. I also include in the animation meshes with lines instead of dots: the moiré pattern remains essentially the same, as expected.

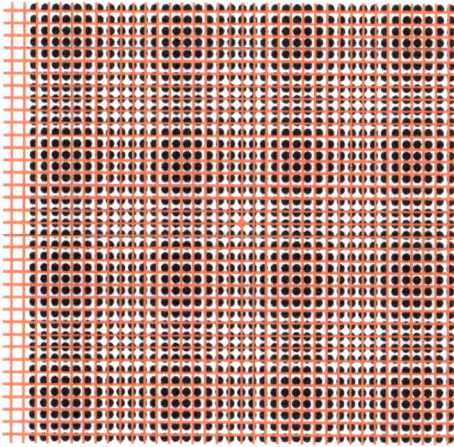

Animation 9-A4 screenshot: The animation is available for download (*see* footnote 1 in this Chapter).

What happens if we add a third mesh, as in the usual situation of sheer curtains? The outcome is hard to predict, but we may wonder whether it will be anything like the three-fences case of Figure 9-11. Two examples are shown in Figure 9-13, one with dots and a different one with lines. In this relatively simple case, we have three square meshes with different sizes.

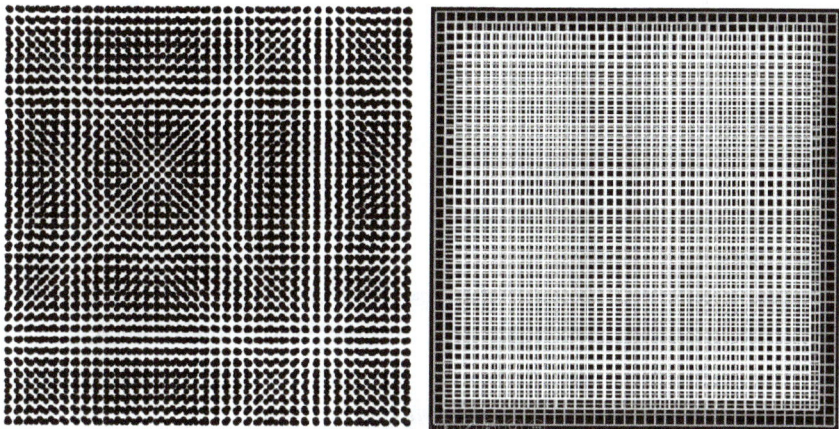

Figure 9-13: At left, three square meshes of dots (all black) produce a complex moiré pattern. At right, we have a set of three square meshes of lines (all bright on a black background). The meshes have different sizes. The right pattern mimics looking out from a home at night through three layers of a sheer white curtain which is illuminated from inside.

As in the 1D case of 3 fences, the resulting 2D wave pattern is rather wild, but even more complex: there is no repeating pattern. This moiré pattern is also very sensitive and mobile, as you may see in Animation 9-A5.

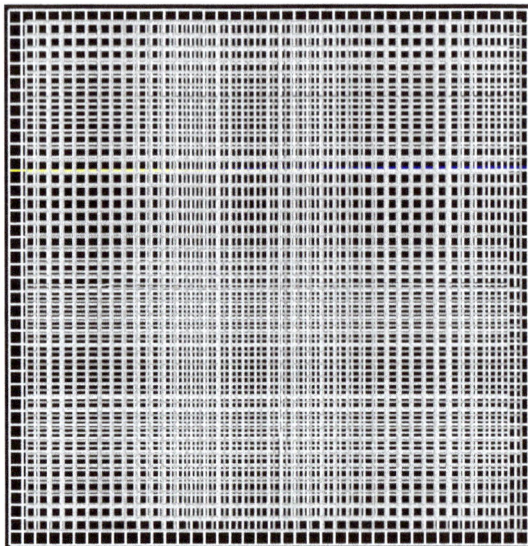

Animation 9-A5 screenshot: The animation is available for download (*see* footnote 1 in this Chapter).

Two dimensions offer an important new opportunity: the relative rotation of meshes. What can we expect? An example is shown in Figure 9-14. Here a square mesh is rotated by 7 degrees, without changing its size. The rotated and the unrotated meshes produce a wavy moiré pattern; as before, this pattern is totally artificial, since it does not exist in the individual meshes.

Note again the similarity between the pattern obtained with lines (at left) and that obtained with dots (at right): the same wave shape and size are produced. Also notable is that the moiré pattern is itself rotated relative to both meshes: thus, the wave direction is rotated as well.

You may view the very dynamic effect of a rotating 2D mesh in Animation 9-A6. Similar to the case of growing and shrinking meshes (Animation 9-A4), we here observe: **The wavelength of the moiré pattern is large when the two meshes are closely aligned; the wave-**

length shrinks rapidly when the two meshes are rotated. When the two meshes are not closely aligned, the wavelength becomes so small, comparable to the mesh spacing, that there is no clear wavy pattern. Here again, I include in the animation meshes with lines instead of dots: the moiré pattern still remains the same.

Figure 9-14: A square mesh of lines (at left) or dots (at right) is rotated by 7 degrees, forming a moiré pattern against the unrotated mesh.

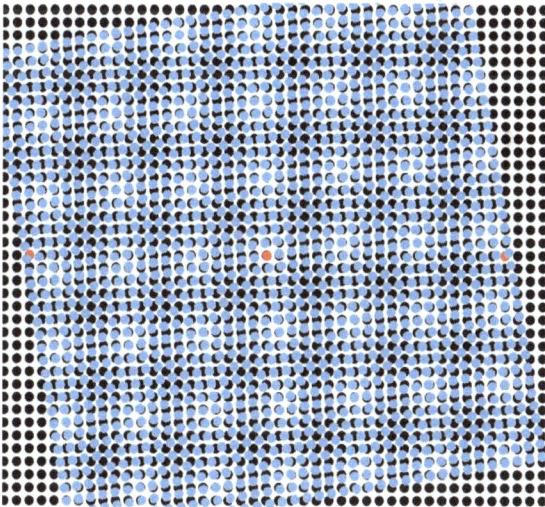

Animation 9-A6 screenshot: The animation is available for download (*see* footnote 1 in this Chapter).

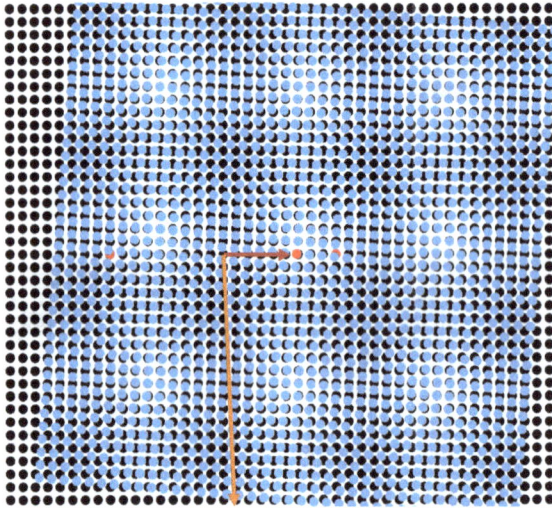

Animation 9-A7 screenshot: the animation is available for download (*see* footnote 1 in this Chapter).

What happens when a rotated 2D mesh slides over a non-rotated 2D mesh? Surprise: the moiré pattern slides in a different direction! This is exhibited in Animation 9-A7. As indicated in the image shown next, shifting the blue mesh to the right (along the red arrow) causes a shift nearly downward (along the orange arrow) of the moiré pattern. As you can imagine, adding a third layer (as in sheer curtains) makes the pattern even more complex; since this case does not teach us more, I do not illustrate it here.

You may be curious to know why the moiré waves in Animation 9-A7 move in unexpected directions, and also why they move quite fast: to see pictorial explanations, watch Animation 9-A8, illustrated below: its multiple frames exhibit the mechanism of this surprising behavior.

We see that 2D adds several surprising "twists" to the behavior of wavy moiré patterns, thanks to the possibility to rotate meshes.

You will find several beautiful animations of other moiré patterns entitled "Moirons at work" by Klaus Hermann on YouTube.[4] These

---

[4] https://www.youtube.com/watch?v=IPojeIUcO58,
https://www.youtube.com/watch?v=njN2HUa-vBw,
https://www.youtube.com/watch?v=7m2MNkjAobs, and
https://www.youtube.com/watch?v=bzSmhCY7Ybg.

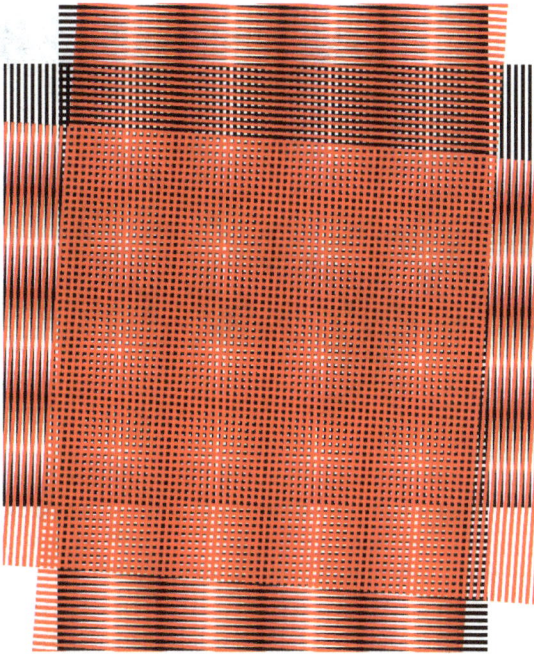

Animation 9-A8 screenshot: The animation is available for download (*see* footnote 1 in this Chapter).

represent atomic structures at crystal surfaces. One such structure is labelled "graphene overlayer shifting": it shows a single layer of graphite lying on another single layer of graphite (graphite is pure carbon and is commonly used as the "lead" in pencils); the same structure is also shown here in Figure 9-15.

A single layer of graphite is called **graphene**. The "bilayer of graphene" has become very famous in physics, because when two layers of graphene are rotated by about 1.1 degrees from each other, this material suddenly becomes superconducting: this means that the pair of graphene layers loses all resistance to electrical conduction (along the layers) and therefore can carry electrical currents without loss. That would be wonderful for carrying electricity over large distances from power plants to customers. Unfortunately, this superconductivity exists only when the graphene is cooled to a temperature of a few degrees above absolute zero (which is −273.15 degrees centigrade). The problem is that it is extremely expensive to cool an electrical grid to such extremely cold temperatures!

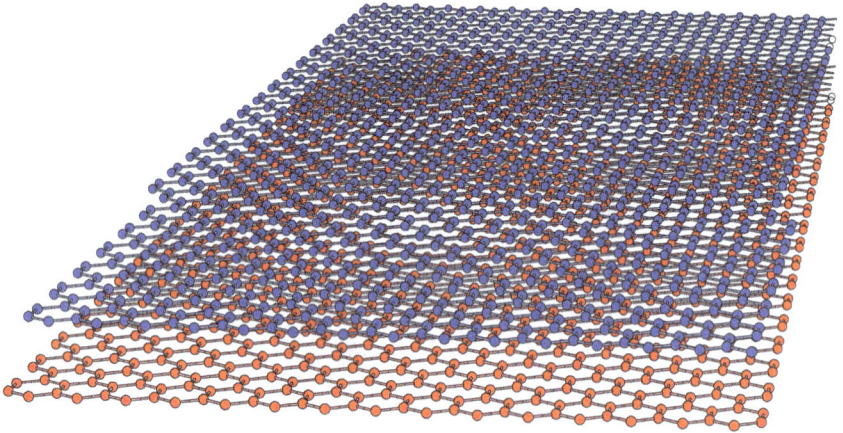

Figure 9-15: This oblique view shows pieces of two layers (blue and red) of "graphene": each layer is composed of a honeycomb lattice of carbon atoms (shown as small balls linked by lines). The two layers are here rotated by 1.1 degrees relative to each other: this creates a wavy moiré pattern visible where the two layers overlap. The two pieces of graphene in this image are much too small to see by the human eye: their size is roughly a millionth of the thickness of human hair; actual pieces of bilayer graphene can be larger than a millimeter and therefore visible, but the atomic-scale details shown here remain far too small for the naked eye. (*Source*: Drawn by the author with K.E. Hermann's Balsac software.)

Not only does "bilayer graphene" with 1.1-degree rotation exhibit superconductivity, but it also produces a wavy moiré pattern, as seen in Figure 9-15.

You may still wonder: ***In Figures 9-1 and 9-5, why are the 2D moiré waves themselves so curved?*** In this Section, we have seen several examples in which the waves of the moiré patterns are quite straight, for instance in Figures 9-12 and 9-14. The difference between curved and straight waves comes from two effects. One is perspective: whether you or the camera looks perpendicularly at the meshes (along their surface normal). This perspective effect is shown in Figure 9-16: we get curved waves that form a moiré pattern with the display pixels.

The curves in the waves seen in Section 9.1 also come from another effect: the use of real *versus* idealized meshes. In this Section we have used mathematically precise square meshes, while the examples in Section 9.1 have meshes that are themselves somewhat curved or irregular. The sheer curtains and chair back nets in Figure 9-1 are indeed

Figure 9-16: This perspective view looks at a flat rectangle seen "off-normal" by a few degrees, like looking at a painting on a wall from off-center. The rectangle contains only "horizontal" black lines (that are parallel to the "horizontal" top and bottom edges), on a yellow background. In this 2D view the "horizontal" lines form a fan of lines broadening out to the right, causing a wavy moiré pattern by interplay with the display pixels. By magnifying the image on your screen, you may see more and stronger moiré waves.

not perfectly flat and also can be stretched or rotated (twisted) unevenly, resulting in local changes in the relationship between the meshes. Interestingly, the very curved waves on the display screens shown in Figure 9-5 imply tiny irregularities in the alignment of display pixels in the displays or of detector pixels in the camera (or both); "tiny" here means comparable to the size of the pixels themselves, and so essentially undetectable by the naked eye; these moiré patterns can therefore be used to test the perfection of the pixel arrays in displays and cameras.

You may also wonder: ***How do <u>colored</u> moiré patterns arise, like those in Figure 9-5?*** Remember that those patterns were observed by viewing a uniform white electronic display with a digital camera. This means that two meshes are involved: the mesh of emitter pixels in the display, and the mesh of detector pixels in the camera. Both meshes have red, green and blue sub-pixels. Basically, the color seen by a sensor sub-pixel depends on which sub-pixel of the display it sees.

A simpler model that also exhibits this colorful moiré effect is shown in Figure 9-17. A mesh of red, green and blue dots plays the role of the white-emitting sub-pixels in the display; however, this model has 2 green dots for each red dot and each blue dot, so we have a mesh

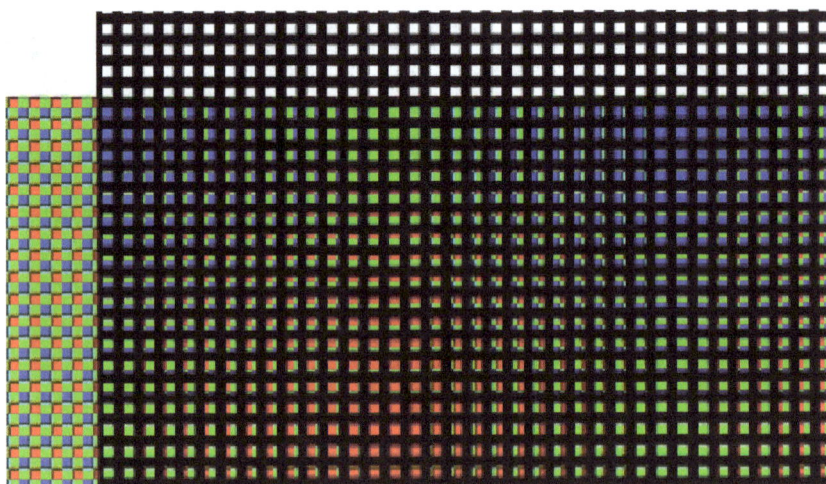

Figure 9-17: A square mesh of red, green and blue dots (seen uncovered at left and identical to the Bayer pattern of Figure 5-17) is covered with a square mesh of lines (shown alone at top). The black mesh has holes that allow seeing a dot of one color, while blocking dots of other colors. The black mesh is shrunk by 3.5% relative to the colored mesh, causing the colorful moiré pattern. The 3.5% shrinkage translates into a nearly 30-fold magnification between the colored dot pattern and the moiré pattern.

dominated by green, as we see at left in the figure. A black mesh with holes plays the role of sensor sub-pixels: these holes are small enough that they allow seeing just one color. By looking closely, you will see that the moiré pattern in this case is a very magnified view of the tiny pattern of colored dots seen at left (this is the Bayer pattern shown in Figure 5-17).

We can show that such colorful moiré patterns also exhibit the complex motions which we saw earlier, but now in color: Animation 9-A9 slides and rotates the black mesh which we discussed above over the same colored mesh; it also combines a red, a green and a blue mesh to produce colors similar to those of Figure 9-5; another example will be shown in Figure 10-3.

A final question: **where do the _pastel_ colors in Figure 9-5 come from?** They indeed look much lighter than the colors seen with my model in Figure 9-16. This comes from my simplified "camera", namely my black mesh: it blocks about 3/4 of the light coming from the model

Animation 9-A9 screenshot: The animation is available for download (*see* footnote 1 in this Chapter).

"display". The blocked light comes from emitter sub-pixels of different colors, so that, if they were included in my model, we would get complex and brighter mixtures of red, green and blue: that would then also produce pastel colors.

9.4 What have we learned in this Chapter?

Wavy moiré patterns are formed by the combination of two or more simple regular meshes, producing waves of regular dark and bright (or colored) bands that did not exist in the original meshes. Such patterns are most familiar in sheer curtains, in photographs of patterned clothes and, thanks to software, to lesser extent in images on computer displays, smartphones and television screens.

Moiré patterns are very sensitive to motion of the meshes that are combined, leading for example to dancing wavy patterns in sheer curtains due to a breeze or our breathing.

One-dimensional moiré patterns are easiest to understand. Two-dimensional patterns add interesting and sometimes surprising effects due to rotation of 2D meshes.

10

Optical Illusions

*Our eyes, together with our brains, frequently play visual games on us, called **optical illusions**. Perhaps the most familiar optical illusion is the stroboscopic or wagon-wheel effect, often observed in movies: a spinning wheel or propeller appears to rotate much more slowly than in reality, or even backwards, or not at all, giving a comical impression. Another type of optical illusion is more subtle: some drawn images look realistic, but lead to "impossible" objects.*

There are many more kinds of optical illusion: many can be viewed on the web. A convenient list is given at Wikipedia.[1] You may also watch optical illusions described in YouTube videos.[2]

---

[1] https://en.wikipedia.org/wiki/List_of_optical_illusions.

[2] https://www.youtube.com/watch?v=78T848QuaME and https://www.youtube.com/watch?v=xYe4-7I5ot0. See also two videos *"Optical illusions in 2D"* and *"Optical illusions in 3D"* on Everyday Physics by Michel A. Van Hove at YouTube: https://youtu.be/V5M29Pz82tU and https://youtu.be/2yBkLs0YDhU.

Here we will select a few optical illusions and categorize them to present a coherent perspective on what they look like and why.

⬤⬤⬤

10.1   Now you see it, now you don't: Time and color

In many optical illusions the image that we see changes over time. For example, the color may appear to change, but often it is only our perception that changes: this effect is called afterimages, described in Section 10.1.1.

Another category of optical illusions relies on recognizing structural patterns that we "add" to the image. For example, facial recognition can identify a single person from very different images of that same person. Our memory contains a face for that person: we use that memorized face as a pattern to identify that person. But we may misidentify the person in an optical illusion. We will give and discuss examples in Section 10.1.2.

A further class of optical illusions relies on differences in color contrast. Shadows and shades are good examples, described in Section 10.1.3.

10.1.1   *Color changes: Afterimages*

A very famous color illusion is the **afterimage**: it results from staring for a minute or so at a picture and then quickly switching to look at a blank surface; you then see something else.

To experience this **optical illusion in two dimensions**, based on **color changes**, look at Figure 10-1. After staring at the "x" in the left image for about a minute without moving your eyes, what do you see after you quickly switch to look at the "x" at right? You should see two half-circles: the top one is yellow, the lower one cyan; both should be brighter than the white page, but both will fade away within a minute or so.

Also do this little experiment: once you see the half-circles at right, tilt your head sideways, to the left and to the right: you will see the half-circles tilt as well. This proves that the effect is in your eyes, not in the drawing, computer or display!

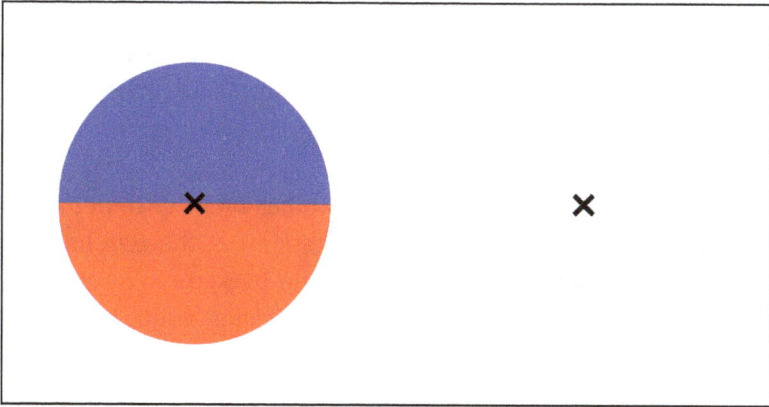

Figure 10-1: Watch the afterimage: stare for about a minute at the left "x", then switch to the right "x": see the new colors around the right "x".

In this illusion, red turns into cyan, while blue turns into yellow. ***Do you remember the relationship between red and cyan, and between blue and yellow?*** Look at the circle of colors in Figure 10-2 (copied from Figure 2-17 in Section 2.7). It is a circle in which colors opposite each other are "complementary". There you see red opposite cyan, and blue opposite yellow. This illusion thus converts a color into its complementary color.

What causes this afterimage? The answer is actually closely connected with the concept of **complementary colors**, as follows. When red light falls on your retina, it excites primarily the "red" cones. These cones thus work hard to produce signals for the **brain**. Working hard results in the cones getting tired: this reduces the strength of the signal to the brain from the "red" cones. Put simply: "getting tired" is the effect of using up the chemicals needed to produce that signal, so those cones become less able to continue producing that signal, just like muscles get tired and weaker when producing a force.

Now, when you suddenly switch to staring at the white area below the "x" on the right in Figure 10-1, the red cones are still tired and give a weaker signal, while they continue receiving plenty of red light coming from the white area. But now the "green" and "blue" cones also get excited by the white light: this initially results in strong green and blue signals (which together give bright cyan) and a weaker red signal in the lower half circle. It's like removing some red from white, which gives precisely the complementary color to red, namely cyan, although with still some residual red, resulting in a whitish cyan, or "pastel cyan off-white". Similarly, in the upper half circle at right, your "blue" cones will be tired, while your "red" and "green" cones will give strong signals, producing yellow.

Why are the cyan and yellow afterimages brighter than the white background? Remember that, while you stared at the two half-circles at left, your eyes were also seeing the white background around these half-circles: looking at white also makes the cones tired, so the white gradually becomes weaker and thus slightly gray (we don't normally notice this, because we are so used to changes in brightness). Therefore, when switching to the right, our eyes see the white background as a bit darker (light gray) than it was initially.

If you continue staring at the lower right-hand area for a minute or so, the "green" and "blue" cones gradually also become tired, so that all cones — red, green and blue — give weaker signals, and the originally white area becomes uniformly very light gray: the afterimage dissolves into a featureless light gray. (Indeed, any white area gradually looks light gray after about a minute, but our pupil and brain compensate very effectively for these changes in apparent brightness.)

The two half-circles that you see on the right are probably not very sharp: do you know why? The reason is: while you were watching the left half-circles, your eyes were twitching, jumping around somewhat, blurring the impact on your cones near the edges of the half-circles. That blurring is also kept in the afterimage, blurring its edges. For this reason also, you can't produce a complementary afterimage of the thin black frame around Figure 10-1: it is too blurred by your eye movements.

Why do we twitch our eyes all the time? The main purpose of twitching the eyes is to better see the edges between regions of different colors or shades, since those are important for identifying what objects we see. Twitching the eyes overlays neighboring regions and thereby highlights their color contrast.

Maybe you have observed the following in Figure 10-1: as you stared at the left half-circles, you may have noticed new colors flashing along their edges, in the shape of thin Moon sickles. These colors are actually the cyan and yellow that you will soon see when looking to the right: they have the same origin, namely that your eyes twitch while the cones at those edges are getting "tired". The sickles tell you that your eyes are ready to switch to the right and see the afterimage!

Now repeat the same viewing with the color circle of Figure 10-2: stare at the color circle at left for about a minute, then quickly switch to the blank area at right. What do you expect to see? Based on our discussion above, each color at left should be replaced at right by its complementary color. Do you see that?

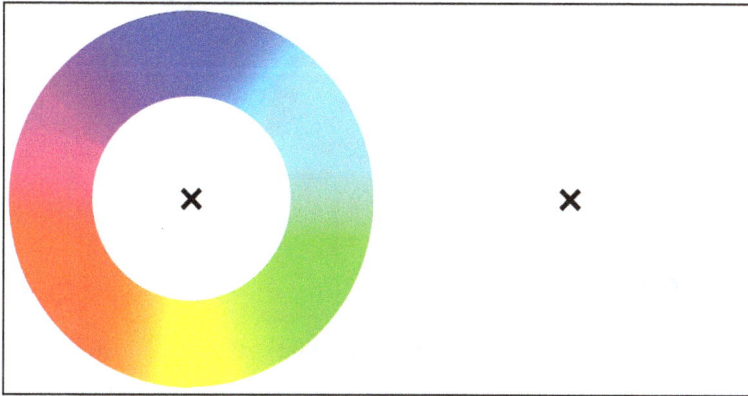

Figure 10-2: Watch the afterimage at right, after staring at the image at left.

As we discussed in <u>Section 2.7</u> (see especially Figure 2-17), the complementary color is exactly opposite in the same circle: this means that the color circle should be inverted (rotated by 180 degrees) when you watch its afterimage. In other words, the red at about 8 o'clock in the figure (bottom left of a clock) should reappear at about 2 o'clock (top right); the blue at about noon should reappear at about 6 o'clock; and similarly for all other colors around the color circle (although again brighter).

You may view an animation of a similar afterimage in Animation 10-A1. That animation is a modification of the so-called "lilac chaser" or "Pac-Man illusion": here white circles sweep around colored rings as your eyes remain fixed. Gradually, your eye cones get tired, so the white circles become colored, adopting the complementary colors of the large circular rings. The result looks like an ice cream cone. But if you move your eyes, the moving circles lose their color and immediately become white again.

Animation 10-A1 screenshot: The animation is available for download.[3]

10.1.2   *Pattern recognition: In time and space*

You must have seen the funny **wagon-wheel effect** in **movies**: a **wheel** or **propeller** that turns rapidly may appear to rotate too slowly or not at all, or it may even seem to incorrectly rotate backward. (Several amazing examples of rotating and "bending" propellers are shown in a YouTube video.[4]) Watch Animation 10-A2 for a simulation of this optical illusion, illustrated below: in the first page of that animation, a wheel rotates slowly clockwise; on successive pages, the wheel rotates faster and faster. This rotation is faithfully shown by the large red arrow, because it is firmly attached to the wheel. But the numerous thin spokes of the wheel appear more confusing. What happens is the following.

The computer display that you may be using to watch this animation rapidly produces individual images one after the other, just like TV screens and movie theatre screens. Most computer and TV displays produce 60 images ("frames") per second. This is fast enough that motion usually appears fluid and continuous to our eyes; our eyes retain

---

[3] Animations are available to buyers of this book at https://worldscientific.com/worldscibooks/10.1142/12316#t=suppl (for more details, see References and Resources on page 257).

[4] https://www.youtube.com/watch?v=xYe4-7I5ot0.

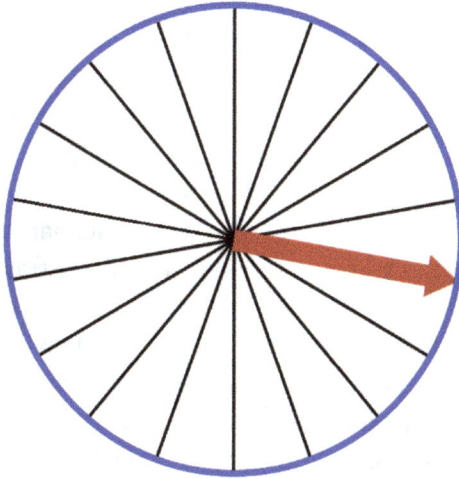

Animation 10-A2 screenshot: The animation is available for download (*see* footnote 3 in this Chapter).

each image for about 1/30 of a second, so that the 60 frames per second refresh the image on our retina before the last image fades away.

Warning: It is possible that your display uses a different "frame rate" or "refresh rate" than 60 frames per second (which is usually written as 60 Hz, where Hz stands for Hertz, a unit of repetition speed). Then the animation will give different effects than those I describe here.

Therefore, when the wheel rotates slowly, as on page 1 of our animation, it appears to move smoothly. But as it speeds up on successive pages, the arrow and spokes jump ahead in rapid, regular jerks: the thick arrow gets blurred and the thin spokes split into multiple spokes.

When we get to page 5 of the animation, the red arrow shows rapid rotation (more than 3 complete revolutions per second). However, the images of the spokes fall just <u>before</u> their images in the preceding frame, making it appear as if the spokes are turning backward, counter-clockwise, and thus opposite the clockwise direction of the actual arrow and wheel rotation. **This is the often-comical optical illusion of backward rotating wheels in movies.**

We next watch page 6 of the animation. The red arrow (and wheel) rotates a little bit faster. But more interestingly, the spokes don't turn at all (at least if your display produces 60 frames per second): **The wheel** appears to have stopped rotating and seems stationary, even though the arrow shows that the wheel actually still rotates fast.

On page 7, the rotation is again a little bit faster: we now see the spokes move slowly forward again (clockwise), while the motion of the arrow looks much the same as before, just a bit faster.

Speeding up further, we again see faster motion of the spokes (page 8), but then once more the spokes appear to stop rotating (page 9): by now the wheel is turning twice as fast as on page 6, but again the spokes seem stationary; nevertheless, the arrow spins around like crazy, making larger steps than before and showing the true speed of the wheel (in fact, at these high speeds, the arrow starts to show the same effects as the spokes did at slower speeds).

If we continued speeding up the wheel, we would see more "stationary states", at speeds that are triple, quadruple, quintuple, *etc.* the apparent "stationary" speed of page 6.

The wagon-wheel optical illusion is very similar to the familiar **stroboscopic effect**, which is much used in shows on stages. A strobo-scope is a lamp that delivers light in short bursts in rapid succession. These fast bursts are equivalent to the frames produced by an electronic display. The stroboscope thereby also produces a rapid succession of images (snapshots), as in movies, and causes very similar optical illusions, such as wheels that apparently rotate backward or seem stationary but actually rotate forward.

The wagon-wheel optical illusion is an example of **pattern recognition**. We look for a stable and familiar **structural pattern** that persists even though the views of that pattern change over time. Here the structural pattern is the wheel: we assume that it is not changed by the rotation. In 2D, simple patterns are lines, circles, triangles, wheels, letters, logos, *etc.* In 3D, simple patterns are balls, combs, cups, forks, chairs, tables, *etc.* We trust in the permanency of such patterns even when their appearance changes: with the wheel, we still believe that it rotates forward faster and faster, even though it seemingly rotates the wrong way or too slowly, including when it seems to stop rotating. The structural pattern contradicts the illusion: if we know the structural pattern (for instance by assuming that the wheel remains normal), we can override the illusion and avoid having to believe in abnormal behavior. A backward rotating or stationary wheel is certainly physically possible, but also very abnormal and improbable given our knowledge of how wheels behave. We are therefore willing to accept that we are fooled into seeing an artificial backward rotation. **This is typical of optical illusions: we have to choose between contradictory interpretations; some interpretations are visually attractive (illusions), while others are more realistic (stable patterns).**

Let's now consider a different optical illusion, which has a different pattern: the stepping motion of feet. Watch Animation 10-A3: this **stepping-feet illusion** mimics the stepping of a bipedal human (with two feet), of a quadrupedal animal (with four feet) and of a centipedal animal (with a hundred or many feet). The feet are simply shown as little colored boxes that slide very smoothly across black and white stripes. The fun part of this optical illusion is that the stripes give the appearance of jerky step-like motion, despite their smooth sliding: this is more clearly visible when the feet reach the uniform gray area.

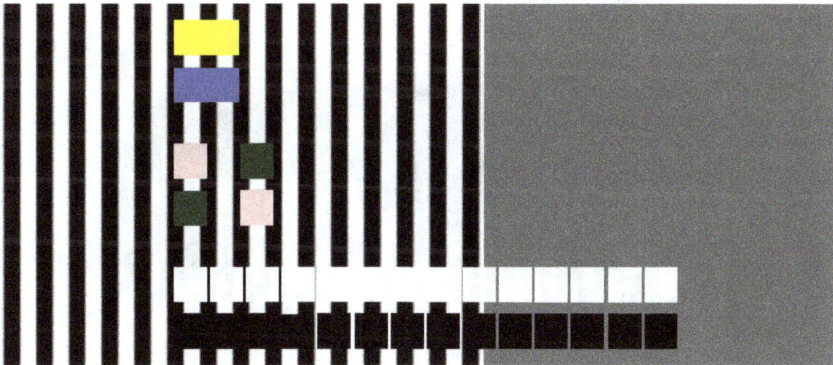

Animation 10-A3 screenshot: The animation is available for download (*see* footnote 3 in this Chapter).

What gives the appearance of jerky step-like motion? The clue is the contrasting colors of adjacent feet. For example, the bright yellow foot is almost invisible in the white stripes, while the dark blue foot is almost invisible in the black stripes. So, the yellow foot's motion is only clearly visible when its ends pass through the black stripes, while when its ends move through the white stripes the foot seems to have stopped moving. This mechanism is even more obvious for the black and white feet.

The stepping-feet illusion reminds us of the wavy **moiré patterns** that we discussed and illustrated in Chapter 9. In Figure 10-3 are two examples of moiré patterns: both have a wavy appearance and are static, without motion; both can be animated for more dramatic effect, as we did in Animation 9-A1 and Animation 9-A9.

Indeed, moiré patterns (discussed in Chapter 9) are similar to optical illusions. In the examples shown here and in Chapter 9, we may ask: *What is the structural pattern and what is the illusion?* The pattern in the image with the comb (top of Figure 10-3) is the repeating structure where the teeth's alignment blocks or transmits light. That alignment is not part of the comb, but due to the relative positions of the comb, its mirror image (and thus the mirror itself) and the camera: in that sense, the wave is an illusion which is foreign to the comb. The pattern does not strictly repeat itself: it varies, even within a single picture, when the angle of view changes, as we saw in Figure 9-6.

Non-repetition in a moiré pattern is most obvious in the colorful pattern in the bottom image of Figure 10-3. It is very hard to recognize

Figure 10-3: A natural moiré pattern (between a comb and its mirror image, at top), and a computer-generated color moiré pattern (composed of three rotated meshes of red, green and blue dots, at bottom).

and describe what exactly the pattern is here: we do see similar colors appearing in different places, in the form of waves with a characteristic length scale and brightness range (there are no sharp wave tops, or very short or very long waves). How would you describe the specific character of this pattern of colored dots that makes it different from almost every other picture you have ever seen? And what is the illusion here? We could interpret this wavy pattern is several ways: it could represent a cloudy sky at sunset, or a mix of liquids with different colors, or reflections off the surface of water. So you have to make a choice between such possibilities to reconcile what you see with something real, even though none of those choices is exactly correct: for example, what is the meaning of the dots?

The reason why I stress the concepts of structural patterns and illusions is that we struggle with this absolutely constantly as we interpret what we see in daily life: Whatever scene we view, we try to explain it in terms of familiar objects (structural patterns), even though there could be other explanations (illusions).

Facial recognition is a clear example (see also Section 5.1.5). We try to extract a fixed and well-defined pattern (the identity of a given person) from an image that can vary a lot (with lighting, angle of view, hair style, glasses, mood, shaving, *etc.*) and therefore can be misleading. The structural pattern is an abstraction (such as the identity of a specific person), while the illusion is an actual image (such as a photograph that looks like someone else, even though it shows that specific person).

Another example of structural patterns and illusions is the case of **illusory contours**. I have no doubt that you can come up with a simple interpretation for the top row of images in Figure 10-4. **What is the pattern?** The structural pattern is white circles over black squares. **What is the illusion?** The illusion is the extension from disconnected bits of contours (bits of circles and bits of squares) to make three full imaginary circles and three full imaginary squares that are not completely present in the image. You could imagine many other interpretations of these three images (such as an unusually white full Moon seen against a black sky through a white window frame from slightly different directions and distances, or white pancakes cut by squares, *etc.*), but most interpretations would be more complex or abnormal: we prefer simple, normal explanations.

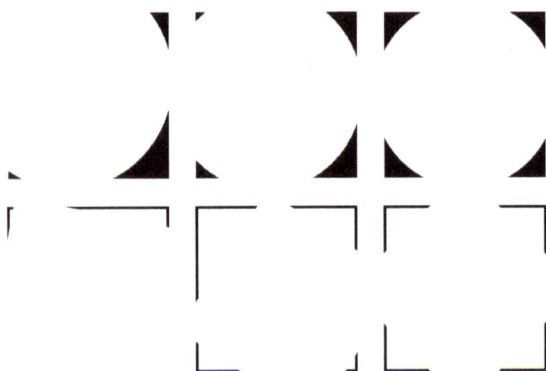

Figure 10-4: Illusory contours: What simple shapes do you recognize in these six drawings?

What do you make of the <u>bottom</u> row of images in Figure 10-4? This I offer as an exercise: find the common pattern and imagine alternative explanations.

Figure 10-5 gives a couple more challenges of pattern recognition, just for entertainment, with unusual fonts (the word "font" means letter style or typeface). Looking at the top 2 lines, written with the Parchment font, what does this sentence say? And in the bottom line, written with a modified Onyx font, what are the letters at left? How did I modify this Onyx font?

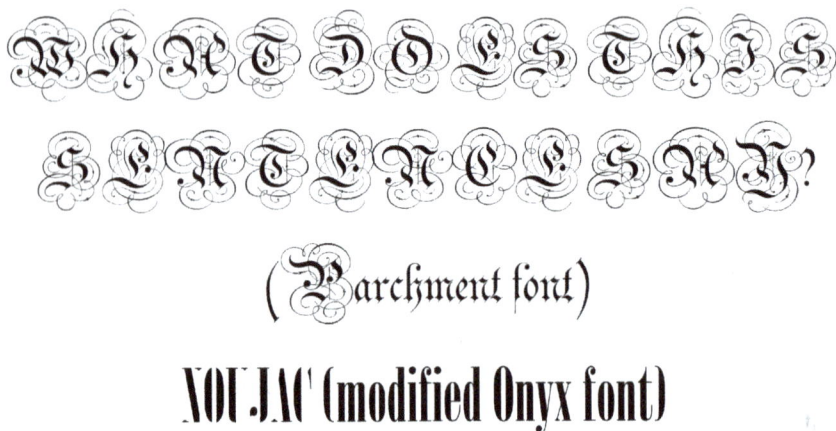

(Parchment font)

XOU JAC (modified Onyx font)

Figure 10-5: What does the top phrase say? And what are the first 6 letters in the last line?

10.1.3 *Contrasting shades*

Let's look at the top pair of boxes in Figure 10-6. Do you think that the small square at right is darker than the small square at left?

When we move the two small squares together, as at bottom in Figure 10-6, we find that they have the same gray level. Indeed, that is how I chose their gray levels. This example illustrates a common **optical illusion** involving **contrasting shades**: **our brain compares an object's observed intensity to that of the surrounding scene, and increases the apparent contrast in shading.**

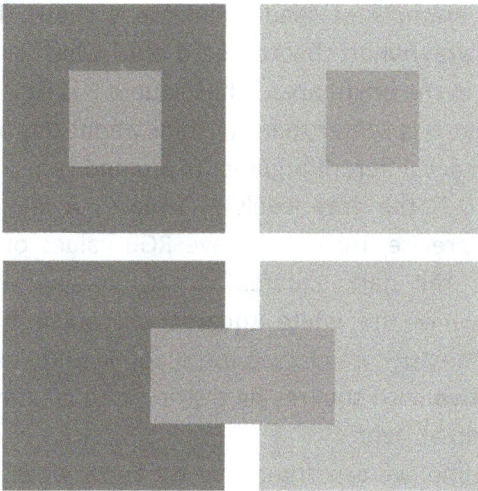

Figure 10-6: Top pair of boxes: Which of the two smaller squares is darker? Bottom: the two smaller squares have been moved together.

This increase in contrast is not simply a change of the pupil size. You can convince yourself of that by switching rapidly between the left and right images, faster than the pupils can react (the pupils take a few seconds to adjust, as you can test by switching bright lights on and off).

Thus, it is the brain (or at least the nervous system) that is increasing the contrast. A stronger contrast is certainly beneficial for interpreting a scene, since it is very useful to detect edges of objects and boundaries between objects to identify their shapes and positions. So, this effect probably has an evolutionary origin and is "hard-wired" into our nervous system.

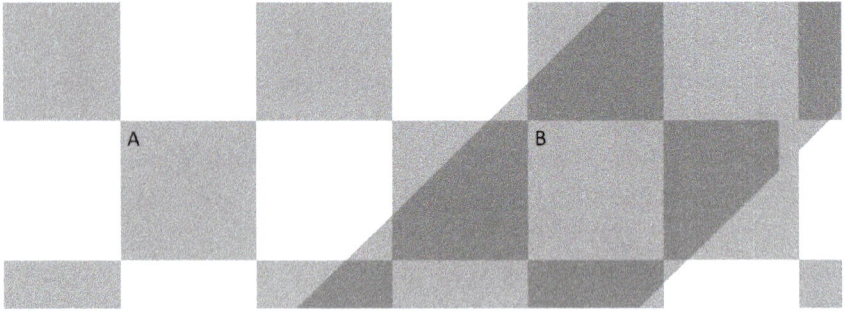

Figure 10-7: A checkerboard crossed by a shadow: is square A darker than square B?

A stronger example of shade contrast is shown in Figure 10-7. Here we have a gray/white checkerboard onto which a shadow is cast. Square A is dark in the bright area, while square B is light in the shadow. Do you agree that B is lighter than A? Think about this question before reading on: consider different arguments pro and con.

In Figure 10-7, the gray levels of squares A and B are actually identical: to be precise, they both have RGB values of (167,167,167); for comparison, the dark squares in the shadow have gray level (123,123,123), while the white squares of course have gray level (255,255,255). An easier way to see the shade identity of squares A and B is to connect two such squares by a strip of equal gray level: see the short vertical strip at right.

In this example, we can argue that we know what a checkerboard looks like: squares with only two distinct colors. That implies that square B must be lighter than square A, despite the shadow that darkens B. However, this argument injects two external pieces of information into the interpretation: the two-color structure of checkerboards and the darkening function of shadows; this information is not available in the image itself, which could well be a faithful image of an unusual checkerboard (for example, a checkerboard that has been painted exactly as shown). In other words, the brain is changing the observed image to simplify it to a more familiar situation: in effect, the brain is compensating for the shadow by automatically "dividing it out". The eye itself cannot do this across a whole image. Cameras also cannot divide out a shadow; but computer processing of images can remove shadows, like the brain.

Thus, we again find a convenient and familiar structural pattern: a normal checkerboard, overlaid by a normal shadow. Our brain automatically removes the shadow to arrive at a simple interpretation, even if it contradicts the visible illusion.

You can find more such optical illusions dealing with contrasting shades on the web.[5]

10.2   Optical illusions in two dimensions

All the images we see are inherently two-dimensional, in the sense that whatever we see is projected onto the 2D surface of our retina. We will discuss the more complex case of three-dimensional viewing in Section 10.3, which involves interpreting 2D images into 3D scenes and structures.

The preceding optical illusions (Section 10.1.1) are all two-dimensional: they mainly involve color and time-dependent effects. There are also optical illusions in 2D that play more with the shapes and sizes of objects: as we will see, some appear to bend lines, while others sow confusion about the size of objects.

10.2.1   *Two-dimensional directional illusions*

Consider the next three illustrations: Figures 10-8, 10-9 and 10-10. They show pairs of colored lines against a set of non-parallel lines or arrow-heads.

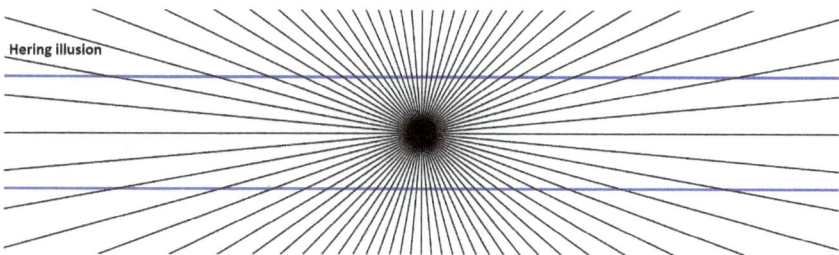

Hering illusion

Figure 10-8: An example of the Hering illusion: are the two blue lines straight or curved? A ruler (or the edge of the page) will tell you!

---

[5] http://web.mit.edu/persci/gaz/# (requires Adobe Flash Player).

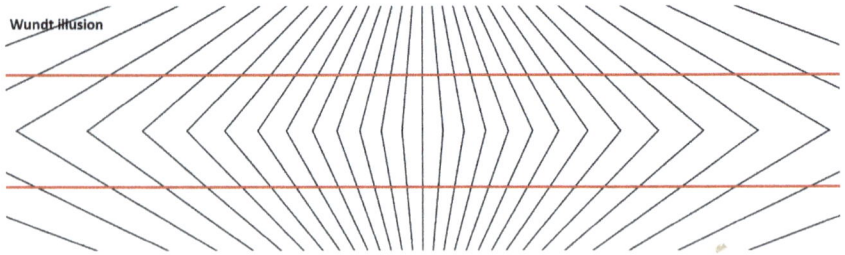

Figure 10-9: An example of the Wundt illusion: are the two red lines straight or curved? If curved, do they match the curving in Figure 10-8?

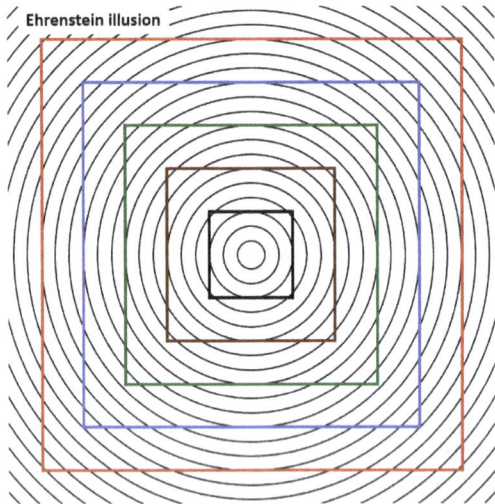

Figure 10-10: Circles instead of radial lines produce the Ehrenstein illusion: are the colored lines straight or curved? If they are curved, in which direction, and does the direction match the Hering and Wundt illusions?

The **Hering** and **Wundt illusions** are classic (Figures 10-8 and 10-9): most people see the straight colored lines as curved, in opposite directions.

If we use a background of circles instead of radial lines, we get another, closely related optical illusion, called **Ehrenstein illusion**, shown in Figure 10-10.

In Figure 10-11, illustrating the **parallelism illusion**, we show how the tilt of the background lines influences the perceived orientation of straight lines. The gray background lines are straight, and all tilted by the same angle. The brain gives the impression that the colored lines in these illusions are more perpendicular to the background lines than is actually the case. A number of explanations have been proposed for this set of illusions, but there is no consensus on which explanation is correct. Presumably, natural evolution has favored "exaggerating" the perceived angles between crossing lines, resulting in the appearance of curves and non-parallelism.

In Figure 10-12, you can let your imagination go wild: what is straight, what is tilted, relative to what?

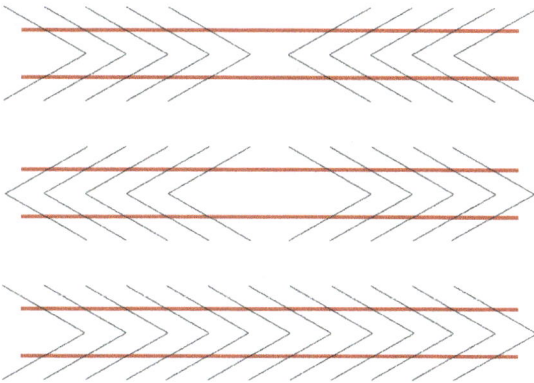

Parallelism illusion

Figure 10-11: Parallelism illusions similar to the Hering and Wundt illusions: are the red lines straight or curved, parallel or convergent?

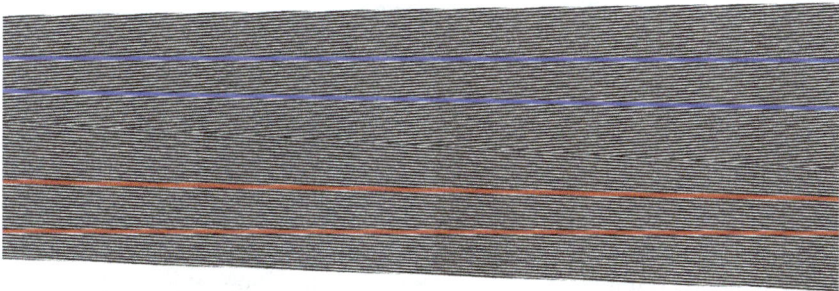

Figure 10-12: More line illusions: using your eyes only, do the colored lines appear straight, zigzag-shaped or curved? Are they parallel to each other and/or to the page edge? (In displaying this figure, you may see mobile wavy moiré patterns, depending on the magnification of the page; moiré patterns are discussed in Chapter 9.)

10.2.2   *Two-dimensional size illusions*

We now turn to optical illusions that confuse our size perception, causing **size illusions**. In Figure 10-13, we pairwise compare the lengths of two line segments. In the **Müller-Lyer illusion**, we see the top blue line as clearly longer than the bottom blue line, despite the very obvious dashed red lines that point out that both blue lines are equally long. In the Sander and Ponzo illusions, the two dotted green lines have equal lengths, while the two red lines also have equal lengths.

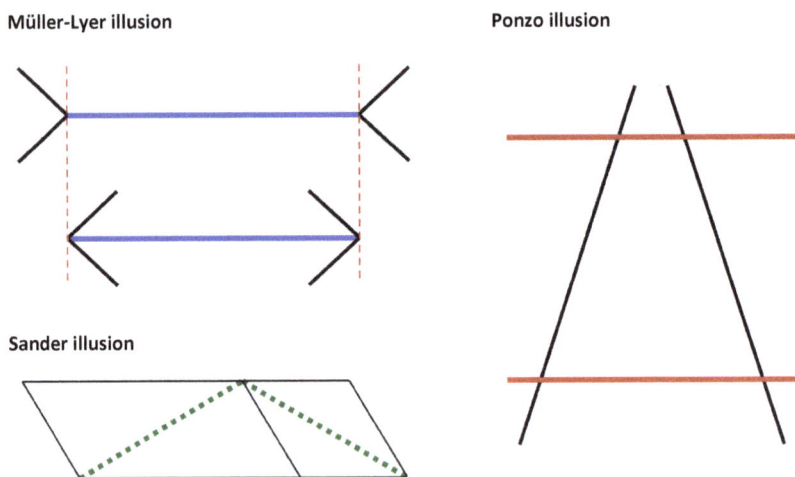

Figure 10-13: In these three optical illusions, we compare the lengths of the pairs of blue, green and red line segments: which of the two is longer in each pair?

The **Ponzo illusion** is the easiest to explain: the two black lines suggest a road (or track) that recedes into the distance. The most familiar and easy interpretation of this (three-dimensional) imaginary scene is that the red object slides away along the road, so that it should appear to shrink with distance, due to the perspective effect. But, since the red line visually does not shrink with distance, we interpret it to become longer, while still sliding horizontally along the road. A very reasonable alternative scenario is that the red line simply rises vertically above the road, keeping the same distance and thus keeping the same apparent size. But it seems that our brain still prefers the more familiar horizontal motion suggested by the road.

Turning to the **Müller-Lyer illusion**, one possible explanation (among others) of an apparent length difference goes as follows. The brain looks more at the <u>bulk</u> of the two black arrowheads than at the <u>end points</u> of the blue lines; the centers of gravity of the two arrowheads are definitely farther apart in the top image (pointing toward each other) than in the bottom image (pointing away from each other).

The **Sander illusion** may be explained in the same manner: the black "arrowheads" formed at the ends of the two green line segments are on average more distant in the leftmost case, which has more widely opened arrowheads, compared to the tighter arrowheads of the rightmost case. Another, very plausible explanation could be that the left-hand green line appears larger because it is in a larger "box" than the right-hand green line. Can you think of other explanations?

Another size comparison occurs in the **Ebbinghaus illusion** shown in Figure 10-14. Here we see again that the perceived size of an object is much affected by the scene surrounding it. A circle looks smaller when it is surrounded by larger circles than when it is surrounded by smaller circles, due to the contrast in sizes. Less apparent in our illustration is that the width of the annular white space between the central circle and its neighbors also affects the comparison: a tighter space makes the central circle look larger. A similar effect is seen with the apparent size of the Moon: when the Moon appears close to the horizon (or close to objects like trees or buildings), it seems much larger than when the Moon is far from other objects in the sky.

**Ebbinghaus illusion,
a.k.a. Lipps illusion, Titchener circles**

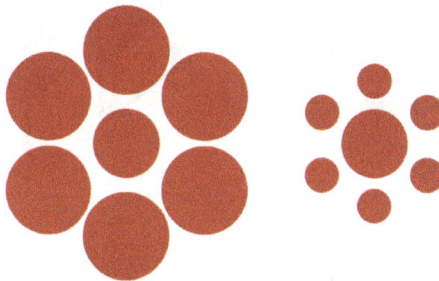

Figure 10-14: This optical illusion, with several names, compares the central circle surrounded at left by larger circles, and at right by smaller circles: is one of the two central circles larger than the other?

You will find many more two-dimensional line illusions on the web.[6]

10.3   Optical illusions in three dimensions

Consider the triangle shown at left in Figure 10-15. This drawing represents a three-dimensional object made of three colored bars which complete a closed triangle. The little cubes simply highlight the 3D shape and help show what parts hide each other.

Now follow the three bars from cube to cube all around the triangle: do you agree that the cubes appear to stick together face-to-face in a perfectly reasonable way?

Next, think of the red (rightmost) bar being vertical, while the other two are both horizontal and perpendicular to each other. Start at the nearest green cube and "walk" clockwise along the three bars.

What happens when you reach the top of the red bar? If you assume that the red bar contains an elevator, you can take that elevator down from the blue bar to the green bar, and thus back to your starting point. But aren't the blue and green bars at the same level? If so, coming down brought you to the same level from which you started, a clear contradiction.

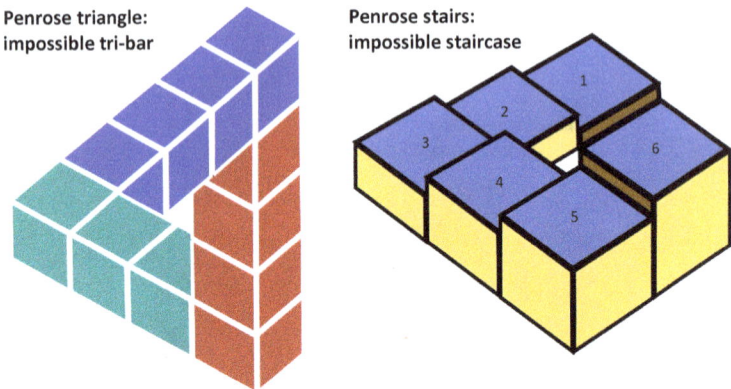

Figure 10-15: Two "impossible" structures: the Penrose triangle (left) and the Penrose stairs (right). The steps are numbered 1–6.

---

[6]For example: http://brainden.com/line-illusions.htm.

Let's repeat this in more detail: Start walking horizontally from the nearest green cube to the left end of the green bar. Make a right turn onto the blue bar and walk horizontally to its end. So far you have been walking horizontally at what we can call "ground level". Next drop vertically down along the red bar (as in an elevator) to its bottom end: since you are going down, you end up well below the green and blue bars, underground. However, big surprise: the bottom end of the red bar connects back to the green bar, as if you had reached ground level from above!

How can you drop underground in the red elevator, and end up on the ground level from where you started?

The answer is: you can't, and this triangle is impossible. Hence this triangle, which was devised by Penrose and therefore called the **Penrose triangle**, is also known as the **impossible tri-bar**.

Another way to convince yourself that the Penrose triangle is un-realistic is to try to build it with cubes, for example wood cubes, sugar cubes or ice cubes: you will not be able to complete the triangle as drawn, whether you glue the cubes together or not (even with floating cubes in water in case you blame gravity for the difficulty)!

The Penrose triangle illustrates the concept of structural pattern that I mentioned in Section 10.1.2. Here we see a pattern that we are familiar with: a bar made of cubes. We believe this bar to be straight and connected perpendicularly to the other bars. Using such bars, our brain produces a simple model of connected bars that unfortunately conflicts with reality, so we have an optical illusion. Before trying to explain this, we can consider another example.

Next look at the looping staircase drawn at right in Figure 10-15. The blue squares are horizontal steps. Imagine standing on the rightmost blue step, numbered 6. Now climb onto step 1 and turn left. Continue by going up two steps, 2 and 3, to reach the left corner of the staircase. There, again turn left and climb two more steps, 4 and 5, reaching the corner at bottom in the drawing. Make another left turn and climb another step. Big surprise again: you reach the rightmost step 6 again, which was your starting point, even though you have been climbing a total of 6 steps, without ever descending! And, as with the triangle, all the connections between the steps look correct. Do you agree?

How can this be? ***How can you climb steps and end up at the same level from where you started?***

The answer is again: you can't, and these stairs are impossible. These stairs, which are very closely related to the Penrose triangle, are called **Penrose stairs** or the **impossible staircase**.

In the image of the stairs, we detect a familiar structural pattern of connected steps: we thus build a **mental model** of stairs, even though they end up going round and round without climbing or dropping. Our model thereby clashes with reality: we again have an optical illusion.

Where does the impossibility of the Penrose triangle and stairs come from? The cause is our mental reconstruction of three-dimensional objects from two-dimensional images, which can cause optical illusions, including the appearance of impossible objects.

We therefore need to think more about how we interpret 2D images as 3D objects. This is very important, because we do it all the time in our daily life as we look around us. And it is vital as our lives depend on knowing what is around us.

Once we better understand 3D vision, we will discuss the triangle and stairs in more detail in Sections 10.3.4 and 10.3.5, respectively. You will even be able to build your own 3D models to examine with your hands what exactly is going on in three dimensions with these "impossible" structures!

What is the origin of the Penrose triangle and stairs? Both the impossible tri-bar and the impossible staircase were designed and popularized by mathematical physicist Roger Penrose and his psychiatrist father. They published their ideas in 1958.[7] Unknown to them, the triangle was actually created much earlier by Swedish artist Oscar Reutersvärd.

Roger Penrose was initially inspired by drawings of the famous Dutch artist **M.C. Escher**, who had produced a number of "impossible" scenes in which perspective played tricks on the brain's interpretation of his drawings. Penrose decided to see what other "impossible" structures he could invent, and came up with the triangle (tri-bar) and the closely related stairs (staircase).

---

[7]L.S. Penrose and R. Penrose, *British Journal of Psychology*, volume 49, issue 1, pages 31–33, 1958, https://doi.org/10.1111/j.2044-8295.1958.tb00634.x.

In turn, Escher was then inspired by the Penrose structures to produce remarkable drawings based on those new impossible structures. Especially beautiful and relevant to our story here are Escher's drawings called *"Waterfall"*[8] and *"Ascending and descending"*.[9]

Escher's wide range of geometrically fascinating drawings were published in book form.[10] That book, including Escher's drawings, is also available online and for free as a PDF file.[11]

The Escher drawings are highly entertaining and thought-provoking: I highly recommend them. They are further discussed in a book by Bruno Ernst.[12] The same author also published an interesting compilation and explanation of many other optical illusions.[13]

10.3.1 *From 3D real scene to 2D physical image to 3D mental model*

When we look at a 3D real scene, two important functions take place, as sketched in Figure 10-16: (1) a 2D physical image of the 3D real scene is formed on the retina of our eyes; and (2) from that 2D retinal image our brain constructs a 3D mental model of the original 3D real scene and displays it to us in our mind.

We may call step 1 **imaging** and step 2 **imagining**. I use the word "imagining" on purpose to indicate that the brain plays a very large role in constructing the 3D model that we see in our mind: indeed, the third dimension must be imagined since it is not present in the 2D image.

A very important difference exists between these two functions: in step 1 **depth information** is lost, while in step 2 depth information is added. (If we use a camera, step 1 includes two more sub-steps: recording the image and displaying it in front of our eyes as if it were the original scene; however, the following arguments remain the same.)

---

[8] See also https://en.wikipedia.org/wiki/Waterfall_(M._C._Escher).

[9] See also https://en.wikipedia.org/wiki/Ascending_and_Descending.

[10] M.C. Escher, *"The Graphic Work"*, Benedikt Taschen Verlag, 2007.

[11] https://www.pdfdrive.com/mc-escher-the-graphic-work-e185883412.html.

[12] Bruno Ernst, *"The Magic Mirror of M.C. Escher"*, Benedikt Taschen Verlag, 2007.

[13] Bruno Ernst, *"Optical Illusions"*, Benedikt Taschen Verlag, 1992 and 1996.

From 3D scene to 2D image to 3D model

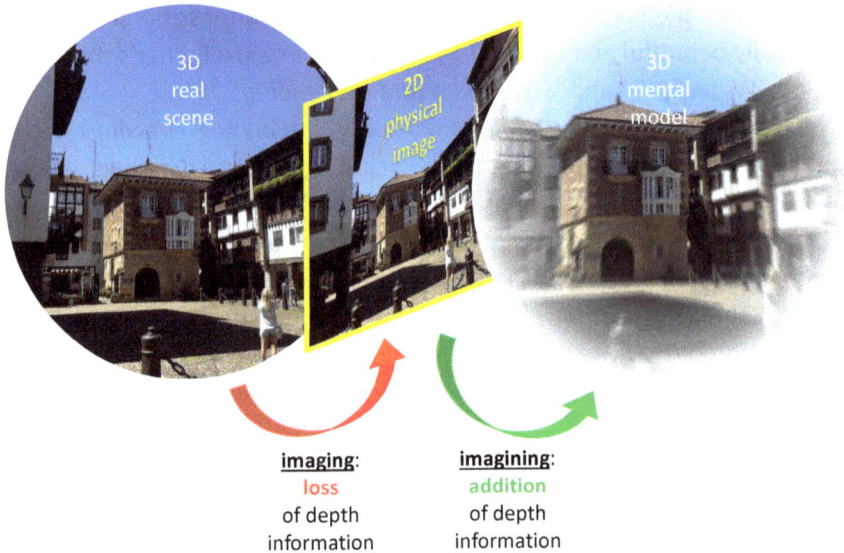

imaging:
loss
of depth
information

imagining:
addition
of depth
information

Figure 10-16: When we look at a 3D real scene (left), a 2D physical image is formed on our retina (center), which is then converted by our brain into a 3D mental image of the 3D real scene (right). In this example, the mental model is basically the same as the real scene (but focusing more on a point of interest, a white balcony): there is no optical illusion, because the brain has received enough information to reconstruct the real scene without confusion.

The loss of depth information in step 1 is inevitable. What we lose is primarily information telling us the distance to objects.

How does the brain add depth information in step 2? This question is much more complex. Depth information is needed to properly understand what you see: that information tells the brain how far different objects are, but that distance information is not directly available in the 2D image. The brain therefore needs to "guess" depth information to properly imagine the placement of the observed objects with respect to each other. The brain performs this complex guesswork by using many clues and much experience. For example, our two eyes see two slightly different views of nearby objects (this is called "**stereoscopic**" vision): the brain uses the differences between those two views to estimate the distances of objects; more precisely, to focus on a closer object, the two eyes squint more toward each other, an effort which the brain translates into shorter distance. The lens adaptation (focusing) of our eyes also gives distance cues. Furthermore, in real life we move, so the view changes: this gives more clues about distance and relative placement of objects. In addition, we know the approximate sizes of familiar objects and thereby also can gauge their distances.

Clearly, the brain performs a complex reconstruction of the original 3D real scene. Normally, the result is as expected: realistic and more or less accurate. That is the case shown in Figure 10-16: the 3D mental model reconstructed by the brain (at right) is essentially the same as the original scene (at left), without contradictions, and therefore without optical illusion.

However, in such a complex reconstruction, occasionally an error occurs: the brain may not have enough information to properly reconstruct a realistic and accurate mental model of a part of the scene. The result is an optical illusion: the brain then has constructed a mental model that does not agree with your experience of reality. Usually, the error is easily and quickly fixed by taking a second look, or the error is ignored if it is not important.

Another source of optical illusions is errors or contradictions within the 2D image, which can happen when this 2D image is drawn incorrectly or contains insufficient information: those are the kinds of optical illusions in which we are interested here, including Penrose's triangle and stairs.

A well-known example is the **Ames room illusion** shown in Figure 10-17 (you may watch a video exhibiting this illusion at Wikipedia[14]). The Ames room is misshapen, but, when seen from a particular position (a peephole), it appears to have the familiar rectangular shape of a normal room: that rectangular shape is the familiar and simple structural pattern, which makes our brain believe that the room is normal, not misshapen.

As a result, when you look through the peephole at a person standing in the far corner of this room, your brain thinks that he/she stands in the nearer corner of a normal room and concludes that he/she is much smaller than expected: when the person moves to the left corner, he/she looks much larger, because he/she is then much closer. This illusion comes about because there is not enough information in

---

[14] https://en.wikipedia.org/wiki/Ames_room.

Ames room illusion

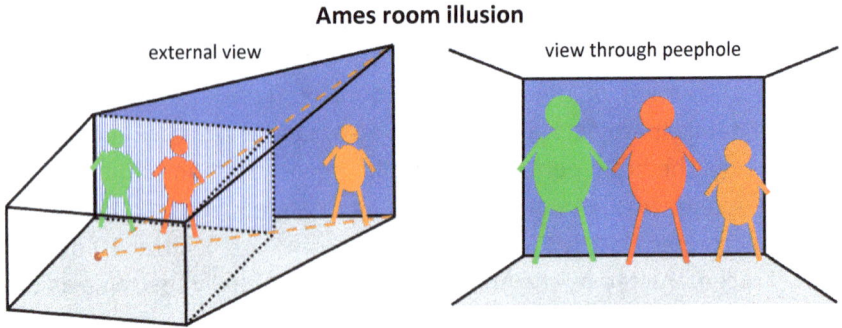

Figure 10-17: For the Ames room illusion, a room is misshapen (left) in such a way that, seen through a peephole in the nearby wall (red point), it appears to have a normal familiar shape (seen at right): the dark-blue wall looks identical to the non-existent light-blue wall, so the viewer thinks the room has a normal rectangular shape. The dashed orange lines show how the right back corner of the room is expanded as seen through the peephole. Therefore, a person standing in front of the dark-blue wall (orange) appears to be much farther away, and therefore much smaller, than when standing in front of the non-existent wall at left (green or red).

the 2D image to show the true misshapen structure of the room: there is no clue that the room has an abnormal shape, so the brain assumes it has a normal shape.

The Ames illusion results from the **perspective** effect, whereby an object looks smaller when it is more distant. In the normal point perspective of our eyes, more distant objects appear smaller, as also shown at left in Figure 10-18. If we draw the same object with parallel perspective (as commonly done by architects and engineers), distance does not affect the actual drawn size: see the right image in Figure 10-18. However, our brain still unconsciously may correct the image as if there were size reduction with distance, and artificially enlarge the more distant objects in our mental image. In the figure, the more distant square and balls probably look larger to you than the same square and balls in front.

Finally, a remark on **holography**: You may know that optical holograms can be used to produce three-dimensional images (some credit cards include a hologram, mainly to prevent forgery). Such holography also uses a 2D physical image, but distance information is embedded in it (in the form of wave **interference** that causes it to shine like a

Cube in perspective

point perspective   parallel perspective

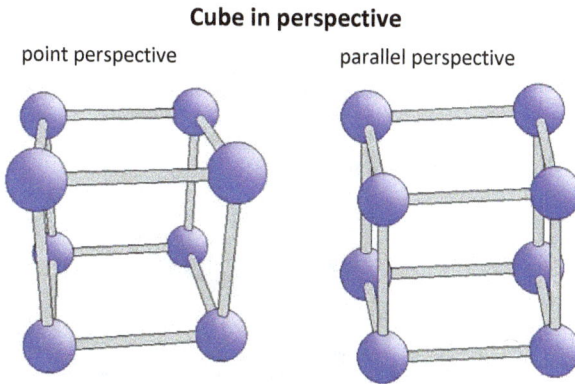

Figure 10-18: A three-dimensional cube is viewed in point perspective (left) *versus* parallel perspective (right), with equal-sized balls at each corner. Do the distant balls at right look larger than those in front? (*Source*: Courtesy of K.E. Hermann.)

solar spectrum). So the distance information is not lost in a hologram: that allows a more faithful reconstruction of the original scene. Unfortunately, holography has some limitations which explain why we do not use it much. One limitation is that it works best with a single color (wavelength), best produced by a laser. Also, the distance information contained in the 2D image is somewhat ambiguous and degrades the image quality. For that and other reasons, only small scenes and images are practical in holography.

10.3.2 *Weaving*

To better understand 3D imagining by the brain, let's consider a simple **weaving pattern** (Figure 10-19: this example will teach us a very important lesson!)

Look at the simple pattern in the top part of Figure 10-19: do you think that we could weave straight threads in this pattern, for example rigid metal wires? Or do we need flexible threads that can snake up and down as shown in the side view in the middle sketch of Figure 10-19? Suppose we make the blue threads straight, as in the bottom sketch, can we also make the red threads straight? The answer is clearly that we cannot weave straight threads, even flat ones: at least some of the threads have to be flexible so they can snake up and down over and under each other.

Weave pattern

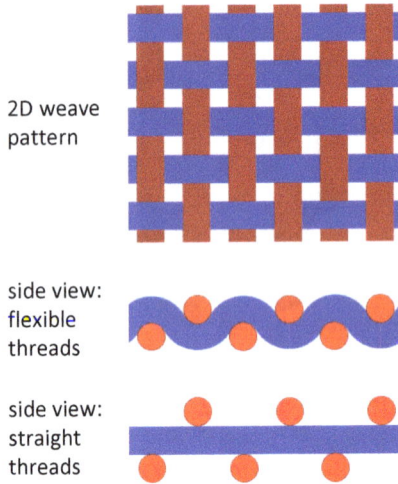

2D weave
pattern

side view:
flexible
threads

side view:
straight
threads

Figure 10-19: Top: weaving pattern composed of threads going over and under each other at each crossing. Center: side view showing how the threads snake up and down over and under each other. Bottom: if we want the blue threads to be straight, the red threads must snake up and down even more.

Why is this important? **The reason is that the plan view of the weaving pattern (top of Figure 10-19) does not give a clue about the structure in the perpendicular direction, the third dimension:** it does not tell us that threads must snake up and down. In other words: there is no information in the 2D weaving pattern about the third dimension.

Naively, we could say: the 2D weaving pattern has simple straight threads, so let's assume that the threads are straight <u>also in the third dimension</u>. After all, we want to keep it as simple as possible. But we see that this leads to an impossible weave: there is no way to make straight threads weave as shown! In the plan view, there is no problem. But as soon as we try to draw side views with straight threads, we produce an impossible structure: an optical illusion.

This is the root cause of many optical illusions: the assumption that straight lines in a 2D image must also be straight lines in the corresponding 3D structure. We will see this again and again. (Several impressive examples are shown in a YouTube video.[15])

---

[15] https://www.youtube.com/watch?v=xYe4-7I5ot0.

Note that the straight threads here are the **structural pattern**. Here we wish to use the simplest possible straight-thread pattern to build the 3D structure: and that is what leads to trouble.

With this weaving example in mind, let's look back at the "impossible" **Penrose triangle** of Figure 10-15. There we see what appear to be simple straight bars; at least they are straight in 2D, just like the weaving pattern. So, for simplicity, we could assume that the bars are also straight in the third dimension, as we tried for the weaving pattern. That is precisely where we run into trouble! As it turns out, when we try to connect them to each other, the straight bars cannot make a proper 3D structure: we thus get an optical illusion. (The straight bars are the structural pattern in the case of the Penrose triangle: we try to build the 3D model using that pattern of straight bars.)

By analogy with the weaving pattern that requires flexible threads to produce a realistic weave, what do you think could be an alternative to the straight bars in the Penrose triangle?

The simplest answer is to allow the bars in the Penrose triangle to curve (snake) or bend in the third dimension, like the snaking threads in weaving. For example, the red bar could twist itself from behind the blue bar to the front of the green bar by making a wild S-shaped curve; this would be possible with a red bar made of flexible rubber, but only if the "painting" of cubes on the red bar produces the fake appearance that it is not curved. Clearly, this can only be done correctly if viewed from one very particular direction, chosen such that we don't see the snaking (just as the simple straight weave is only correct if viewed perpendicular to its plane so we don't see the threads snaking). We will look at this situation more closely in Section 10.3.4.

The case of the **Penrose stairs** is similar. Here, the familiar structural pattern is the square blue steps with the brown risers that connect them. We wish each step to be connected to the next by one riser, as appears in Figure 10-15. The technical problem occurs between step 6 and step 1: that riser seems to go from level 6 to level 1 (counting step heights as levels), which is actually a drop of five levels rather than a rise of one level. Thus, that riser is unrealistic! *Can you think of a way to "fake" that riser from level 6 to level 1 into looking realistic?*

One solution, in analogy with the weave and triangle solutions, is to allow that riser to stretch and bend, so it hangs like a drape from level 6 down to level 1. Again this requires viewing from one particular direction only. We will look at this more closely in Section 10.3.5.

We may summarize our important finding of this Section: Optical illusions can arise if we impose too simple structural patterns on the 3D structure that we mentally build from the 2D image. To resolve such optical illusions requires relaxing the pattern, for example by allowing threads, bars and stair risers to actually curve or bend in three dimensions, even though they appear to be straight from the 2D image.

We will apply this lesson about reconstructing the third dimension to several optical illusions in the Sections 10.3.3 to 10.3.5.

10.3.3   *Concave-convex*

We here consider other optical illusions that depend directly on deciding the up-*versus*-down direction in the third dimension: for example, an object could be pointing at you (the observer) or away from you. This choice can make a shape look either pointed toward you, called **convex**, or hollow, called **concave**.

Look at the image at top center in Figure 10-20, which shows the **Necker cube**. *Can you tell which of the two corners A and B is closest to you in 3D?*

For most people, when they stare at that cube, the perception changes over time: sometimes corner A looks closest, at other times corner B looks closest, or neither looks closest; the first two options are shown at top right and top left, and also below. At right, corner A points toward you, so A is a convex corner of the cube. At left, it is corner B that points toward you, making B a convex corner.

It seems that our brain is constantly re-interpreting the image and arriving at different conclusions at different times. The reason is simple: the image does not contain enough information about the third dimension to favor one choice over the other, so the brain tries both of them out, even subconsciously.

Notice again that we have assumed a structural pattern, namely a cube in 3D. The 2D image does not imply that the lines represent a cube in 3D, but we like to interpret the image with that familiar and simple pattern.

Necker cube illusion

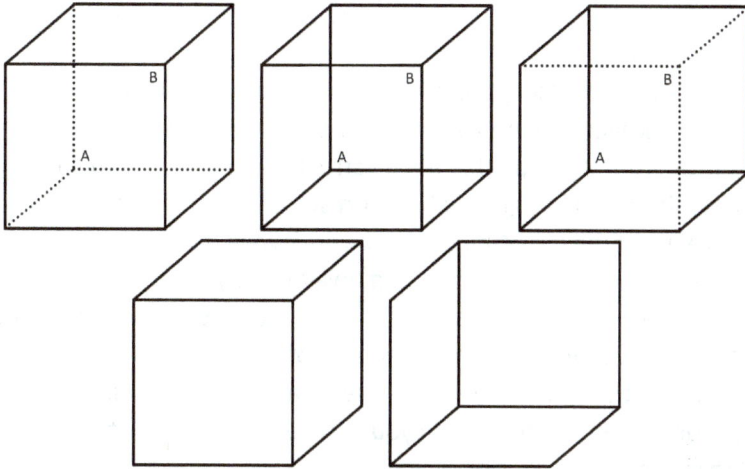

Figure 10-20: Top center: a transparent cube is drawn with all its edges. Top left and bottom left: corner A is imagined to be <u>behind</u> corner B. Top right and bottom right: corner A is imagined to be <u>in front of</u> corner B. The two bottom views, which use opaque faces, appear taken from very different angles, even though the lines have the same directions.

Next look at the two pyramidal shapes in Figure 10-21. ***Do any of these pyramids point toward you (forming a tip) or away from you (forming a pit)?*** What is the reason for your choice between pyramidal **tip or pit**?

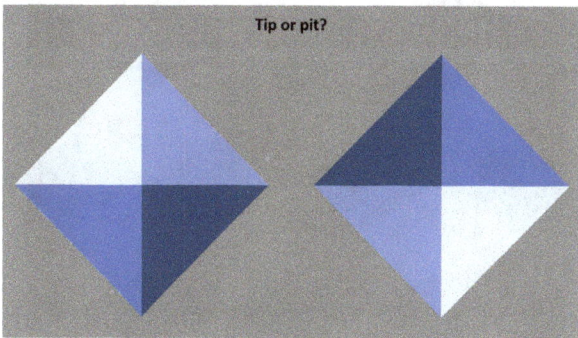

Tip or pit?

Figure 10-21: These two pyramids could point out of or into the page, making them a convex tip *versus* a concave pit. Which pyramid points into the page, and which points out of the page?

Most people probably would decide that the left pyramid is convex, pointing toward you as a tip, while the right pyramid is concave, pointing away from you as a pit.

The image by itself provides no clue to reach this or any other decision. However, we are used to light coming from above, either from the Sun or from a lamp (this is our structural pattern in this case). So the pyramid which is lighter above presumably catches such light and therefore must be pointing toward you, while the other pyramid is lighter below and therefore pointing away from you.

This example illustrates one of the many ways we can decide about the convex *versus* concave shape of objects, based on our own experience with light. However, there is a danger: the light does not always come from "above": then you may reach the opposite conclusion, and create an optical illusion.

Another example of the same situation is shown in Figure 10-22. You are invited to test your interpretive skills on this related pair of images.

Figure 10-22: What 3D geometrical shapes and orientations do you see in these two images, which are identical but rotated?

Based on these principles, the artist M.C. Escher drew a wonderful lithograph entitled *"Convex and Concave"*.[16] It illustrates in a masterful way the conflicts — and opportunities! — arising from this optical illusion.

10.3.4   *Penrose triangle: Impossible tri-bar*

We can now assemble the arguments relevant to understanding the **Penrose triangle**, the **impossible tri-bar**. This triangle is shown again at left in Figure 10-23.

Penrose triangle vs. Model 1 and Model 2

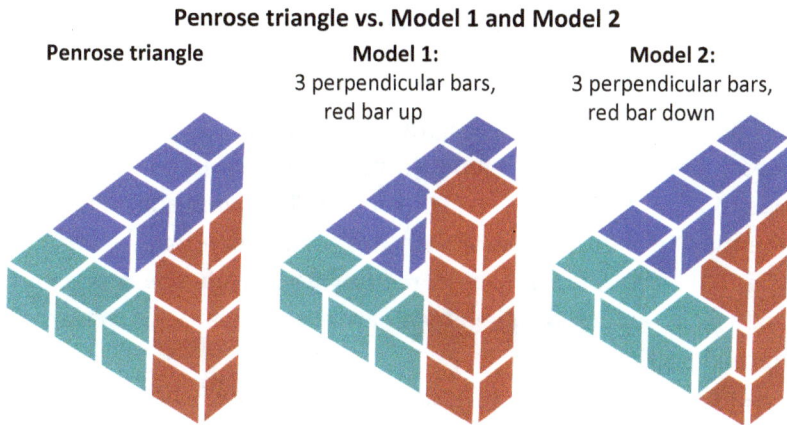

Figure 10-23: Left: The Penrose triangle, as in Figure 10-15. Center: Model 1 with 3 perpendicular bars, the red bar pointing up in front. Right: Model 2, the same as Model 1 except for the red bar pointing down in the back. Instructions for constructing Model 1 are given in a PDF file available for download (*see* footnote 3 in this Chapter).

How do we interpret a 2D image? Our typical mental process when looking at a 2D image goes as follows:

1) From the 2D image we extract a simple, realistic and familiar structural pattern. In this case (at left in Figure 10-23), the pattern is a set of 3D straight bars (made of connected cubes) that connect at their ends with right angles.

2) Using this pattern of connected bars, we mentally assemble (imagine) a 3D model that uses our knowledge of how such a

---

[16]See https://en.wikipedia.org/wiki/Convex_and_Concave.

structure fits together in the real world, while closely matching the 2D image. In this case, we wish the 3D bars to remain straight and connect at right angles.

3) The foregoing two steps are done automatically and unconsciously by the brain. Usually the resulting mental model is successful, and our brain can turn its attention to something else.

4) However, if we consciously consider our mental model more carefully, we may discover contradictions with other facts that we know but ignored in making the model. Now we have an optical illusion: our mental model clashes in some way with known reality. In this case, the straight 3D bars cannot connect as drawn in 3D.

It is important to be aware of the risky logical jump in this mental process: It is dangerous to jump from parallel straight lines in the 2D image to straight lines in the 3D structure. This jump adds information in the missing third dimension. But this assumption may be wrong, as the 3D lines could actually be curved or bent while still projecting to straight lines in the 2D image: we saw exactly that with the weaving pattern in Section 10.3.2.

Let's see how an optical illusion arises with the Penrose triangle. At center and right in Figure 10-23, I have drawn two models using three bars connected at right angles and matching the original 2D image to their left as close as I could. Here you probably can see that the red bar forms either a tower (center image) or an inverted tower (right image). And you also will realize that in both models it is impossible to connect the free end of the red bar to the blue or green bars. But these are perfectly realistic models: you could really build them, although they don't match the Penrose triangle correctly.

It is actually easy to understand geometrically why no model which connects straight bars with right angles can look like the Penrose triangle. In brief: if the three bars form a triangle, they lie in the same plane, so that none can be perpendicular to both of the other two. In more detail: Imagine that the three bars are very thin. Then connect two straight bars, say the green and blue bars: their end points form the corners of a triangle, which defines a single plane. The line connecting the two free ends of the green and blue bars completes the triangle, and so must also lie in that plane. The red bar in the Penrose triangle is supposed to connect those free ends: it therefore must lie in the same plane as the green and blue bars. Therefore the red bar cannot be perpendicular to the green and blue bars.

We have seen in Figure 10-23 two realistic models that can be built for the Penrose triangle, even though they do not match all expectations. Those two models can be built by gluing cubes to each other, such as dice, wood cubes, sugar cubes or ice cubes (you may also cut wood bars to desired lengths: note that the lengths of the bars may be unequal; the same is true of the number of cubes used for each bar).

Can we think of other realistic models? In the following, we will describe several more models, with instructions on how to build some of them yourself with paper.

Models 1 and 2 of Figure 10-23 rely on being viewed from just the right direction, such that the free end of the red bar merges as closely as possible with the free end of the blue or green bar: see also the 3D paper model at left in Figure 10-24. But the mismatch is still very visible: can we hide that mismatch better?

Penrose triangle vs. Model 1 and Model 3

Figure 10-24: Left: photograph of 3D paper Model 1. Center: 2D image of the Penrose triangle, as in Figure 10-15. Right: photograph of 3D paper Model 3. At center, the colors highlight the three bars. At left and right, the colors highlight the orientation of faces: green is horizontal, blue and red are vertical faces, relative to the gray supporting surface. Note that Model 1 here has a taller vertical bar than in Figure 10-23: 5 cubes instead of 4 cubes, for comparison. Instructions for constructing these Models 1 and 3 are given in a PDF file available for download (*see* footnote 3 in this Chapter).

Model 3, photographed at right in Figure 10-24, also made of paper, looks very much like the original Penrose triangle, but it must also be viewed from the correct direction. (I give instructions for building this and other models in a PDF file available to buyers of this book for free download. *See* footnote 3 in this Chapter.)

How can Model 3 look so much like the Penrose triangle? The answer is that Model 3 is derived from Model 1 (see Figure 10-23), but the top of the vertical bar there is made to blend into the appearance of the bar behind it. How this is done is shown from different angles of view in Figure 10-25: the model is certainly not what you "imagined"! The top of the vertical bar has a horizontal extension that matches the orientation and shape of the sidewall of the more distant bar below it, without even getting close to it, let alone touching it. Only with the proper viewing angle does the vertical bar seamlessly cover (and hide) the distant bar.

Model 3 may seem extreme: is there no simpler way to mimic the Penrose triangle? Remember that in going from straight lines in the 2D image to lines in the 3D model, we are allowed to bend or curve those lines. That is what we concluded must happen in the weaving example (see Section 10.3.2 and Figure 10-19). Model 4 does that for the Penrose triangle, as shown in Figure 10-26. Here the vertical bar is bent down to mimic being projected onto the flat green surface. Model 4 has a much simpler form than Model 3 and must also be viewed in the same special direction; however, it is less "clean" than Model 3 due to the difficulty of making the bend invisible.

While Model 4 bent the vertical bar, we may also curve that vertical bar (again as for the weaving pattern). This is done in Model 5, shown in Figure 10-27. (I did not make a paper model, but drew it over a picture of paper Model 3, here seen from the right side.) The advantage here is that the "vertical" bar now connects from below to the bar at right: this matches the Penrose triangle perfectly (if viewed from the same special direction). The disadvantages of Model 5 are the complexity of making that curved bar, and of making it look like the vertical straight bar that it is supposed to mimic.

We have now seen several models that can be built for the Penrose triangle and that give the appearance of the original 2D image. Many other models are also possible, although none will satisfy the structural pattern that we imagined: none have three straight bars connecting at right angles in 3D.

Penrose triangle: Model 3
4 views of the same model

Figure 10-25: Four views of the same Model 3 shown in Figure 10-24. (Here, holes are visible in the back side, in the leftmost view: they become invisible when looking from the proper direction, as in the third view from left.)

Penrose triangle: Models 3 vs. 4

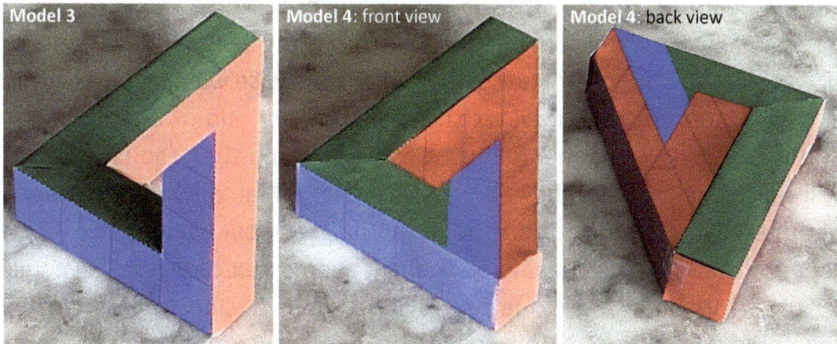

Figure 10-26: Model 3 (left image) is compared directly with Model 4 (center image), which has a flat top instead of a vertical bar. In Model 4, the vertical bar of Model 3 is mimicked by projection onto the green plane. This is done by bending the vertical bar, resulting in the artificial folds seen near the bottom where the vertical bar connects to the horizontal bar. The right image looks at Model 4 from the opposite end and shows that it is indeed flat (the gray face at left is artificial and is invisible in the correct view of the center image). Instructions for constructing Model 4 are given in a PDF file available for download (*see* footnote 3 in this Chapter).

Penrose triangle: Model 5

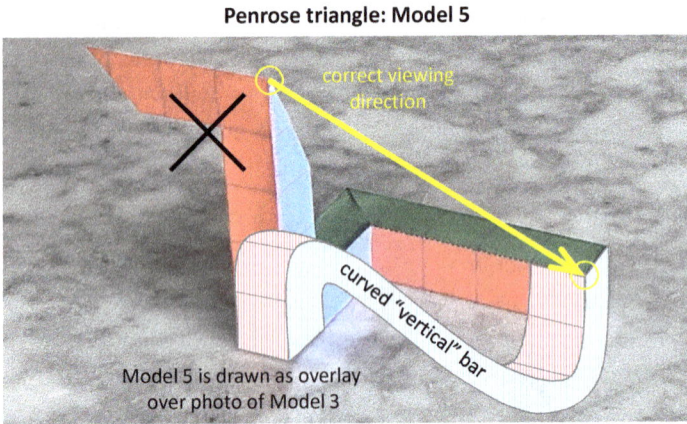

Figure 10-27: Model 5 of the Penrose triangle, overlaid on Model 3 seen from its right side. The "X" on the vertical bar at left indicates that this bar is curved as drawn below. The same viewing direction (yellow arrow) is needed as for Model 3: seen in that direction, the curved bar can look exactly like the straight vertical bar, but it will connect correctly "from below" to the bar at right.

For visual pleasure, I highly recommend viewing M.C. Escher's superb drawing "*Waterfall*": it is derived from the Penrose triangle and gives a beautiful illustration[17] of its puzzling properties. You can indeed easily imagine a waterfall built on the Penrose triangle: in the left image of Figure 10-15, simply replace your earlier clockwise walk with a (slightly sloped) water channel. If water starts at the right end of the green bar and flows along a channel following the green and then the blue bar, it will reach the "top" of the red bar, where it can fall down (in the place of our elevator) to the starting point on the green bar again; there it can repeat the same cycle, again and again. It seems that the waterfall could be used to gain energy (due to gravity) using a waterwheel, and that this could go on forever. This would be a *perpetuum mobile*: a perpetual motion machine that violates basic **laws of physics** by producing energy without energy input.

The Penrose triangle is also described and further illustrated at Wikipedia.[18]

---

[17] https://en.wikipedia.org/wiki/Waterfall_(M._C._Escher).

[18] https://en.wikipedia.org/wiki/Penrose_triangle.

10.3.5 *Penrose stairs: Impossible staircase*

The **Penrose stairs,** illustrated in Figures 10-15 and 10-28, were derived by Penrose from the Penrose triangle. Both have a cyclic structure that "rises" in one direction (or "descends" in the opposite direction), but in the end nevertheless stays at the same level.

Penrose stairs: 2D image vs. 3D paper model

2D image   3D model: 3D model:
                  proper view   side view

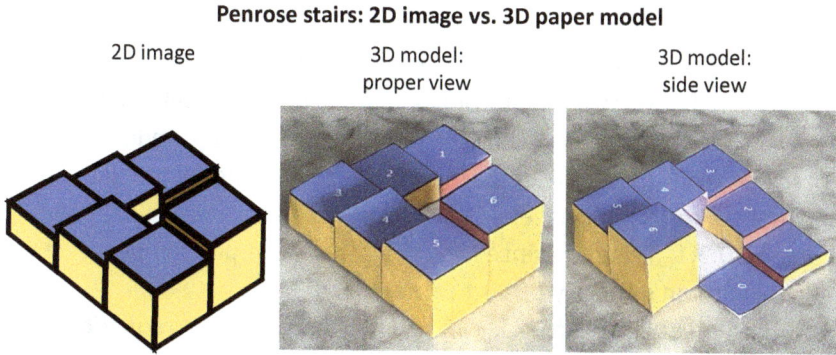

Figure 10-28: Left: 2D image of the Penrose stairs, as in Figure 10-15. Center and right: photographs of a 3D paper model, viewed from the proper direction (center image) and viewed sideways (right image). The numbered horizontal steps are blue, the risers between them brown/red, and the vertical walls yellow (some internal walls are omitted). Step 0 is the ground level. Instructions for constructing this model are given in a PDF file available for download (*see* footnote 3 in this Chapter).

(The Penrose stairs have four sides instead of the three sides of the Penrose triangle, but this is not necessary: the stairs also may have three sides; for example, in Figure 10-28, we can imagine going directly from step 5 to step 1, by eliminating step 6 and adjusting the steps somewhat.)

We can apply the same reasoning to the Penrose stairs as we did for the Penrose triangle in Section 10.3.4. Our mental process will be equivalent, as summarized here:

1) From the 2D image we extract a simple, realistic and familiar **structural pattern**. In this case, the pattern is a set of square horizontal steps that can be connected through vertical risers.
2) Using this pattern of connected steps, we mentally assemble a 3D model. In this case, we wish the steps to remain horizontal and connected through risers all around the cycle.

3) Although usually the resulting mental model is acceptable, ...

4) ... however, in this case, we discover a contradiction with the well-known fact that stairs that rise (or descend) do not lead to the same level. Now we have an optical illusion.

The side view of the model, at right in Figure 10-28, reveals the source of this optical illusion. The top step 6 can hide the ground-level step 0 and appear to connect directly to the bottom of step 1, but only when seen from the proper direction (middle image).

(You may notice a slight apparent size difference between steps 6 and 1 in the "proper view". This is a perspective effect: indeed, step 1 is farther from the camera than step 6, as is visible in the side view, and therefore looks smaller. This is also illustrated in Figure 10-18 in Section 10.3.1. This difference would disappear in parallel perspective, namely with an infinitely distant camera.)

Can we "fix" the staircase model to make it more realistic in 3D, the way we tried for the Penrose triangle with various alternative models? There are at least two possibilities: One option for the stairs is to connect the left back edge of step 6 to the bottom of the riser to step 1: we could stretch a sheet of paper between the two. Another way is somewhat similar to that of Model 5 for the stairs (Figure 10-27): we can curve that sheet of paper from the left back edge of step 6 to the top (instead of the bottom) of the riser from step 0 to step 1.

The Penrose stairs are also described and further illustrated at Wikipedia.[19]

I very much recommend several wonderful drawings by M.C. Escher based on the Penrose stairs, available online: *"Ascending and Descending"*,[20] *"House of Stairs"*,[21] and *"Relativity"*.[22]

---

[19] https://en.wikipedia.org/wiki/Penrose_stairs.

[20] https://en.wikipedia.org/wiki/Ascending_and_Descending.

[21] https://en.wikipedia.org/wiki/House_of_Stairs.

[22] https://en.wikipedia.org/wiki/Relativity_(M._C._Escher).

10.4   What have we learned in this Chapter?

We have systematically discussed a few categories of optical illusions, in order to better understand their appearances and mechanisms. You may wish to view and recognize — or explain! — in online videos[23] some of the optical illusions which we have discussed. Many more varieties of optical illusions are available, as given by links at the beginning and throughout this Chapter.

The first category which we considered is the afterimage, an optical illusion which results from our eyes' cones getting "tired" watching certain colors. This illusion causes us to see complementary colors that are not present in the scene being viewed.

Another category concerns pattern recognition, including the famous wagon-wheel effect seen in movies. It also arises in wavy moiré patterns.

Recognizing structural patterns turns out to be common to many optical illusions: we seem to recognize a familiar structure and use it to build an imagined "reality" that may not be realistic.

Shades also can cause optical illusions, as the brain can change the observed contrast in colors to better identify objects, for example in spite of differences in lighting.

Other geometrical illusions exist in both two and three dimensions. In 2D, we are concerned, for example, with the parallelism of lines and the size of line segments or objects: we often have difficulty with this in daily life!

Optical illusions in 3D are especially interesting. We find that we try to extract from a 2D image some familiar and simple structural patterns, such as rectangular rooms, straight wires and bars, or steps connected as staircases. We then mentally construct an "imagined" 3D model that matches the 2D image. However, every now and then, our imagined 3D model conflicts with reality, producing impossible

---

[23] See, for example, https://www.youtube.com/watch?v=78T848QuaME and https://www.youtube.com/watch?v=xYe4-7I5ot0.

structures such as Ames rooms, Penrose triangles and stairs that violate natural laws.

The source of such illusions is that, in going from a 2D image to a 3D model structure, we have to add depth information that is missing from the 2D image: our brain must guess such depth information, but may guess incorrectly out of ignorance.

Finally, in a PDF file available for download (*see* footnote 3 in this Chapter), you will find instructions for making paper models of "impossible" structures, namely the Penrose triangle and the Penrose stairs: these allow you to get a better feeling for the transition from 2D images to 3D structures.

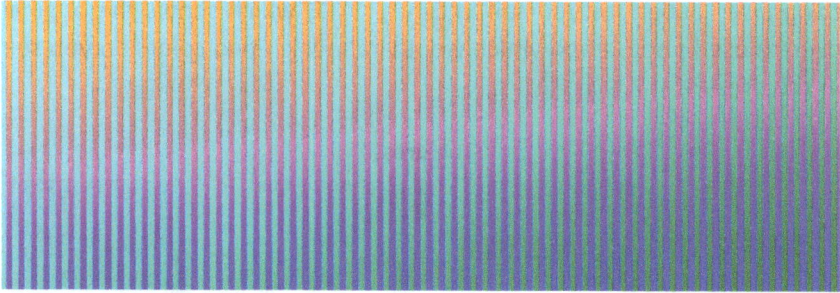

11

Afterthoughts

11.1 From colors to "controlled hallucinations"?

In the preceding Chapters, we have explored vision starting from simple **colors** (red, green, blue) and ending with creative imagination (optical illusions). This exploration continues today: recent research even calls our normal vision "**controlled hallucination**" and suggests that it is related to our **consciousness**. This may sound provocative, but let's think more about it.

In 2015, a photograph of a woman's dress caused a viral online sensation, because of the simple question: what are the **dress' colors**? The surprise was how insistent people were in their radically different opinions about this dress' colors. For example, a subsequent scientific study[1] polled 1401 people, asking them what colors they thought they saw. The response was: *"57% of subjects described the dress as*

[1]Rosa Lafer-Sousa, Katherine L. Hermann and Bevil R. Conway, *Current Biology*, volume 25, pages R523–R548, June 29, 2015, https://doi.org/10.1016/j.cub.2015.04.053.

blue/black; 30% as white/gold; 11% as blue/brown; and 2% as some-thing else." Judge for yourself in Figure 11-1, which shows the original overexposed photograph which went viral: which two colors do you see in the dress?

"The dress"

| Brightened colors from photo | Colors from photo | Overexposed published photo | Darkened colors from photo |

Figure 11-1: The dress photograph was published strongly overexposed as shown here. To its right is the color interpretation by almost 57% of people polled. At far left is the color interpretation by about 30% of people polled. At middle left are colors extracted from the overexposed photograph. The properly exposed **dress colors** were close to the blue and black shown at right. (*Source*: Cecilia Bleasdale.[2])

The extreme and persistent contrast in people's interpretations of these colors remains unexplained, despite many attempts. However, *"the leading explanation for the differing perceptions of the garment holds that people who spend most of their waking hours in daylight see it as white and gold; night owls, who are mainly exposed to artificial light, see it as blue and black"* (quoted from Anil K. Seth[3]). You may not have expected this explanation!

---

[2] https://en.wikipedia.org/wiki/File:The_Dress_(viral_phenomenon).png.

[3] Anil K. Seth, *"Our Inner Universes"*, *Scientific American*, Sep. 2019, pages 34−41, https://www.scientificamerican.com/magazine/sa/2019/09-01/.

One way to think about this picture is **white balancing**, which we have mentioned several times: the **brain** tries to correct the colors to achieve a more acceptable white, but the question is "What is supposed to be white in this picture?". Depending on what your brain assumes to be white in the picture, you can get quite different results; it is indeed easy to over-correct if you have little reliable information. The situation is very similar to the white balancing by my camera shown in Figure 2-2, which turned white plastic into blue plastic. It is also similar to the shade contrast illusion shown in Figure 10-7, where the brain somehow "removes" the effect of shading. A further aspect of the dress saga is that many people probably were not aware that the photograph was badly overexposed, adding one more layer of confusion.

Another way of putting this is: our brain unconsciously injects information into our image interpretation. This may indeed give surprising results, as we saw repeatedly with optical illusions. Where does this injected information come from? Recent research suggests that it must largely come from our own personal past experiences (see, for example, Anil K. Seth[4]): thereby each person's interpretation becomes individual and subjective.

For example, I myself think the colors of the dress are white and gold. This is an unconscious choice, and I guess that my unconscious reasoning goes as follows: the dress seems to be in a shadow, given the very bright background to the right, so its "real" colors should be brighter when the dress is taken out of the shadow. But someone else may well reason as follows: this picture looks overexposed, so all its colors should be darker, making the dress darker blue and darker gray or black.

Next, consider the pair of examples in Figure 11-2, which I drew with simple colored polygons. ***What is your interpretation of the green drawing at left?*** I think you will immediately reach a single interpretation: trees in a forest. ***What is your interpretation of the red drawing at right?*** I think that you will have much more trouble finding an interpretation for this drawing.

---

[4]See, for example: Anil K. Seth, *"Our Inner Universes"*, *Scientific American*, Sep. 2019, pages 34–41, https://www.scientificamerican.com/magazine/sa/2019/09-01/.

Figure 11-2: Two assemblies of colored shapes.

Why is the green drawing so much easier to interpret than the red one? Of course: the green drawing has familiar components. As we discussed with optical illusions, we look for familiar structural patterns with which we can then construct a 3D model of what the image may represent: the green triangles immediately suggest trees grouped in a forest. On the other hand, the odd red polygons are totally unfamiliar and suggest nothing to me (except perhaps abstract "art"?).

You may recognize in this pair of examples the contrast between classical realistic painting and modern abstract painting. **Abstract art** gives the brain great freedom in imagining varied interpretations and in thinking "out of the box". (I hasten to add that I very much enjoy abstract visual art, but more for the colors and shapes themselves than for any "deeper" interpretation that I could make of them.)

Take yet another pair of examples. Suppose you see an image showing half a pizza: how do you interpret this image? As a half-cut pizza or as a half-hidden pizza? Once you have made your choice for the half pizza, address the next question. Suppose you see an image showing half a person: how do you interpret this image? As a half-cut person or as a half-hidden person? I bet that you have chosen opposing answers: a half-cut pizza and a half-hidden person. Why? No doubt because you are more familiar with half-cut pizzas than half-hidden pizzas, and the reverse for persons. This again illustrates how much of our experience we inject into the interpretation of images.

We also all have had this experience: in a movie, we unexpectedly become aware that a terrible accident is about to happen, so we recoil with fright. We have just used our survival instinct to predict an imaginary but terrifying collision, from 2D images that cannot harm us at all.

You can apply this same idea to anything you see: just look around you and ask how much your "view" is affected by what you already know about your environment. For example, you will no doubt assume that your coffee mug has a back side and a bottom, although you don't see them: there is no need to check that fact by turning the mug around or by looking at it from the top or the back. The chances of someone having cut away the back side or the bottom of your coffee mug are very small indeed, and the risk of concluding such an optical illusion is also tiny.

If we couldn't "trust our eyes" (and our brain!), our lives would become very difficult, full of uncertainties and dangers: life would then feel like a constant exploration of unknown and potentially dangerous territory, done slowly and cautiously.

In Seth's words[5] (with my emphases): "*The reality we experience — the way things seem — is not a direct reflection of what is actually out there. It is a clever construction by the brain, for the brain.*" In particular: "***Colors** are a clever trick that evolution has hit on to help the **brain** keep track of surfaces under changing lighting conditions.*" And: "*In this view, our perceptions come from the inside out just as much as, if not more than, from the outside in. [...] perception emerges as a process of active construction — a **controlled hallucination**, as it has come to be known. [...] normal perception is a controlled form of hallucination.*"

We can understand "*controlled hallucination*" this way: in the traditional sense, the hallucinating brain imagines objects and situations that are <u>not connected</u> with the current reality around us, as in a dream or nightmare; but if hallucination is <u>constrained</u> (unconsciously by the

---

[5] Anil K. Seth, "*Our Inner Universes*", *Scientific American*, Sep. 2019, pages 34–41, https://www.scientificamerican.com/magazine/sa/2019/09-01/.

brain) to fit the observed environment (such as an image we are looking at), we get a constructed model of that environment made with our own structural patterns as building blocks. This constructed model we may call "our" reality. Each person's subjective "reality" therefore differs to some degree from the objective reality and from other people's perceived "realities", due to our varied individual prior experiences. This point of view still relies on the existence of one and only one objective reality.

11.2   From colors to the laws of physics

In this book, we have covered aspects of **physics** (electromagnetism and waves), **materials** (detectors and emitters), **chemistry** (cones and rods), **biology** (neurons and brains), and even **psychology** (imagination and hallucinations). And yet, we have barely mentioned some "hard-core" aspects of physics, in particular the **laws of physics** that govern color and vision.

You may know that physics (similar to other "hard-core" sciences like chemistry and mathematics) has a pyramidal structure. At the top of the physics pyramid, mathematical physics provides rigor and precision. But we don't need **mathematics** to understand physical concepts like color, light, vision, energy, mechanics, *etc.*: in this book we can therefore use the conceptual, non-mathematical approach.

Near the top of the physics pyramid reign a handful of extremely powerful, precise and general laws of physics, such as Newton's laws, Einstein's theories, Schrödinger's equation, Maxwell's equations, and the laws of thermodynamics. These fundamental laws of physics govern several large sub-disciplines of physics: **classical mechanics** and **quantum mechanics**, **gravitation** and **relativity**, **electromagnetism**, **thermodynamics**, and more. Such laws dictate how a wide variety of natural phenomena behave, from mechanical motion and nuclear reactions to light and chemistry, to mention just a few: these form the next level of the pyramid. Another level of the pyramid is inhabited by a multitude of practical applications of these phenomena, such as: prediction of eclipses of the Moon and the Sun, weather modeling and forecasting, nuclear reactors for energy generation, health treatments like CT scans, optical tools like corrective glasses and cameras, chemical reactions producing plastics and clothing.

Which of the laws of physics underlie color and vision? The fundamental aspect of color and vision is **light**, which consists of **electromagnetic waves** that obey **electromagnetic laws**. Electromagnetic waves combine both electrical and magnetic waves that can travel together through the vacuum of space (such as from Sun to Earth) and through transparent gases (the Earth's atmosphere), liquids (water) and solids (glass and ice). The behavior of electromagnetic waves is governed by **Maxwell's equations**, developed by British physicist James Clerk Maxwell in the second half of the 19th century. These equations allow us to calculate very precisely how electromagnetic waves travel and evolve over time. They also form the basis for most of our electric and electronic devices, such as electric motors, electric generators, light sources, electronic transmitters and receivers (radios, TVs, smartphones), computers, *etc.* Maxwell's equations also explain the details of a wide range of natural phenomena beyond color and vision, such as lightning and sparks, mirages and fata morganas, and the flashing of diamonds.

Light being a wave, it behaves in many ways like other kinds of waves, such as water waves, sound waves, waves in musical instruments, radio waves, x-rays, *etc.* Color is connected with the wavelength (or frequency) of light waves, as we have discussed in Sections 2.1 and 2.5, and in Chapters 7 and 8. The theory of waves tells us how waves move and gives us precise relations between wavelength and frequency, both in vacuum and in transparent matter.

The creation of light in sources ranging from Sun to lamps and the detection of light by sensors from eyes to cameras depends on **quantum theory**. The quantum laws govern how energy is transferred to emit light and to absorb light, because light in fact carries packets of energy. We have not needed to discuss this more complex aspect in the present book.

Using the basic laws of physics, we can be very precise. For example, we can say that light travels at the amazingly fast speed of precisely 299,792,458 meters per second in vacuum, which is approximately 299,792 (almost 300 thousand) kilometers per second or 186,322 miles per second. To put that enormous **speed of light** in perspective, it takes light about 1.3 seconds to travel from Moon to Earth (that time depends on the varying distance between Earth and Moon), and about 8.3 minutes from Sun to Earth (their distance also varies). Furthermore, we can say that, in one second, light could circle Earth about 7.48 times (around the equator, with the help of mirrors, and a tiny bit more often over the poles, since the Earth is slightly flattened). Also, in one second, light could bounce back and forth between two parallel mirrors facing each other 10 meters apart an astonishing 29,979,245.8 times in vacuum (almost 30 million times, but about 0.03% fewer times in air).

(The reason why light speed is precisely 299,792,458 meters per second, and not an approximate number as would be normal for a measured quantity, is that the meter itself is internationally defined as the distance covered by light in vacuum in 1/299,792,458 second. This number was chosen in 1983 to closely fit the earlier meter-long rod used as the international standard defining the length of the meter.)

Light travels less fast in transparent matter, especially in denser matter like liquids or solids. In air at the Earth's surface, light travels a tiny bit slower, at about 99.97% of its speed in vacuum. In glass or water, light is slowed down to about 67% of its speed in vacuum, so light could circle Earth in an optical fiber about 5.0 times in one second (in practice it would actually be slower because this light would have to be amplified frequently to overcome inevitable losses in the glass; also the fiber may not be able to follow a circular routing around the Earth).

All that high precision is very nice and impressive, but we simply don't need such precision to understand the nature of light, color, vision, *etc.* as we have described and discussed them in this book. Besides, high precision complicates the story unnecessarily: we then have to specify whether we are talking about light in vacuum or in air (with what composition, pressure and temperature?), and whether the light goes around the earth following the equator or over the poles; those details divert our attention from the essential concepts.

Don't we need to use these laws of physics to understand color and light? The answer is no! And the reason is that we can understand the physical phenomena in terms of concepts alone, without having to write down precise formulas that require mathematical skills. It is the difference between "qualitative understanding" and "quantitative prediction". We can indeed grasp the ideas of color combinations (RGB, CMY, HSL) and color blindness without using formulas and mathematics.

As a simple example, we can easily grasp that magenta is a roughly equal mix of red and blue, which is a <u>qualitative</u> description. This description is based on our feelings rather than exact measurements. But if we want to be precise, we can specify mixing exactly 50% of red and exactly 50% of blue, which is a quantitative description. However, such a quantitative description is only meaningful if we supply a very precise definition of what exactly "red" and "blue" are; otherwise different people will produce different "magentas" (likewise, you can mix 50% of apple juice with 50% of orange juice, but your friends will get the same resulting mix as you do only if they use exactly the same apple juice and orange juice as yours). For "daily understanding", we don't need all that detail and precision.

Furthermore, it is interesting that there is in fact no **"theory" of color perception** that is derived from deeper laws of physics, such as quantum mechanics and Maxwell's equations. This is because color perception is partly a function of the nervous system and brain, which we do not know accurately. Instead, there is a "theory" of color perception that is based on measuring human responses to light as averaged over experiments performed on several humans. Physicists call such a theory a "fit" to experiment, or an **empirical theory** or *ad hoc* **theory**. Such a theory stands alone, without need to be supported by deeper laws.

As an example, we have seen in Figure 6-15 the so-called "**CIE 1931 color space**", which is an official map of visible colors: this concept includes the RGB color triangle that we have used extensively in this book. However, when you dig deeper into what that "CIE 1931 color space" actually is, you find that it is "fit" to human responses obtained in simple experiments with small numbers of people; it is not based on any deeper physical theory or laws. In particular, it uses artificial primary colors (which replace the primary R, G and B colors): these imaginary colors do not exist in reality and cannot even be produced in reality.

This does not make the resulting "CIE 1931 color space" and RGB triangle wrong or useless. On the contrary, this color theory is very practical, and it very successfully forms the basis of all color treatment in electronic environments (digital cameras, computers, television, *etc.*).

Biological science also increasingly uses similar "empirical" theories that are fit to experiments, without being based on deeper physical laws. This is quite natural: biology (and even more so psychology) is too complex to be described precisely by fundamental physical laws. Complex systems deserve their own type of theory that is more appropriate and convenient for describing their much more complicated behavior than the traditional simple and idealized "point masses" and "monochromatic waves" of physics.

In conclusion, I hope that you have enjoyed our journey from colors to hallucinations, focusing on conceptual physics, while avoiding unnecessary mathematics and burdensome precision.

Supplementary Material

The supplementary material is free to buyers of this book. It includes:

(1) Animations in Chapters 9 and 10. The most important animations are also included in the YouTube videos about Everyday Physics by Michel A. Van Hove, listed in References and Resources.

(2) Cutout Models in Chapter 10.

Please follow the instructions below to download the files:

1. Go to https://worldscientific.com/r/12316-supp
2. You will be prompted to login/register an account.
3. Upon successful login/register, you will be redirected to the book's page.
4. Click on the "Supplementary" tab to download the files.

For enquiries, please email: sales@wspc.com.sg.

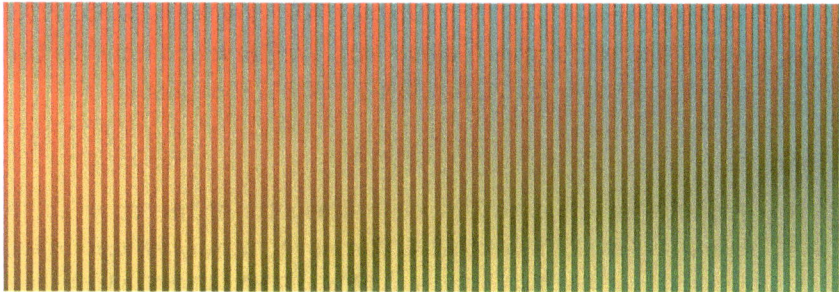

References and Resources

Animations and cutouts pertaining to this book

- A series of animations (in PowerPoint format) and cutout models (in PDF format) exhibiting topics in Chapter 9 (about moiré patterns) and Chapter 10 (about optical illusions) are available for download: *see* Supplementary Material on page 255.

General literature

- Robert G. Greenler, *"Rainbows, Halos, and Glories"*, Cambridge University Press, Cambridge, 1999, ISBN-10: 0521236053.
- Paul G. Hewitt, *"Conceptual Physics"*, Pearson Addison Wesley, 2015 (12th ed.), ISBN-13: 9780135205815.
- Steven Holzner, *"Physics I & II for Dummies"*, Wiley, 2011 (2nd ed.), ISBN-10: 1119293596 & 0470538066.
- M. Minnaert, *"The Nature of Light and Colour in the Open Air"*, Dover, New York, 1954, ISBN-10: 0486201961.
- Richard A. Muller, *"Physics and Technology for Future Presidents: An Introduction to the Essential Physics Every World Leader Needs to Know"*, Princeton University Press, 2010, ISBN-13: 9780691135045.

- Adam Rogers, *"Full Spectrum: How the Science of Color Made Us Modern"*, Houghton Mifflin Harcourt, 2021, ISBN-10: 1328518906.
- Michael M. Woolfson, *"Colour: How We See It and How We Use It"*, World Scientific, 2016, ISBN-13: 9781786340856.

Videos

- YouTube videos exhibiting central aspects of this book about Everyday Physics by Michel A. Van Hove: https://www.youtube.com/playlist?list=PLWIjtByUJvcuf0GCjqGJHNNXMdT1NIsx8.
- University of Wisconsin-Milwaukee, "Science Bag" public lectures, especially those by Robert G. Greenler and by Alan Schwabacher: https://uwm.edu/science-bag/category/all-videos/.
- YouTube videos featuring Neil deGrasse Tyson.
- YouTube video series "Smarter Every Day" by Destin Sandlin: https://www.youtube.com/user/destinws2.
- YouTube video series by "Physics Girl" Dianna Cowern: https://www.youtube.com/c/physicsgirl/videos.

Websites

On **physics**:
- http://hyperphysics.phy-astr.gsu.edu/hbase/index.html

On **color and vision**:
- https://en.wikipedia.org/wiki/Color
- https://en.wikipedia.org/wiki/Visual_perception
- https://en.wikipedia.org/wiki/Color_vision
- https://www.handprint.com/LS/CVS/color.html

On **color blindness and color deficiency**:
- https://en.wikipedia.org/wiki/Color_blindness
- https://www.color-blindness.com/coblis-color-blindness-simulator/
- https://davidmathlogic.com/colorblind/
- https://enchroma.com/pages/test
- https://colorlitelens.com/color-blind-test.html

On **human facial recognition**:
- https://www.scientificamerican.com/article/one-face-one-neuron/

- https://www.scientificamerican.com/article/how-we-save-face-researchers-crack-the-brains-facial-recognition-code/

On **optical illusions**:
- https://en.wikipedia.org/wiki/List_of_optical_illusions
- https://www.youtube.com/watch?v=78T848QuaME
- https://www.youtube.com/watch?v=xYe4-7I5ot0
- https://www.pdfdrive.com/mc-escher-the-graphic-work-e185883412.html

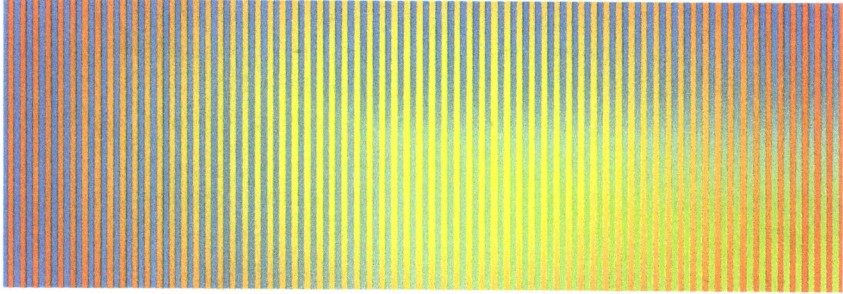

Index

Numbers refer to pages. See also the List of Concepts, Connections and Terminology on page xiii for brief definitions

About the Author

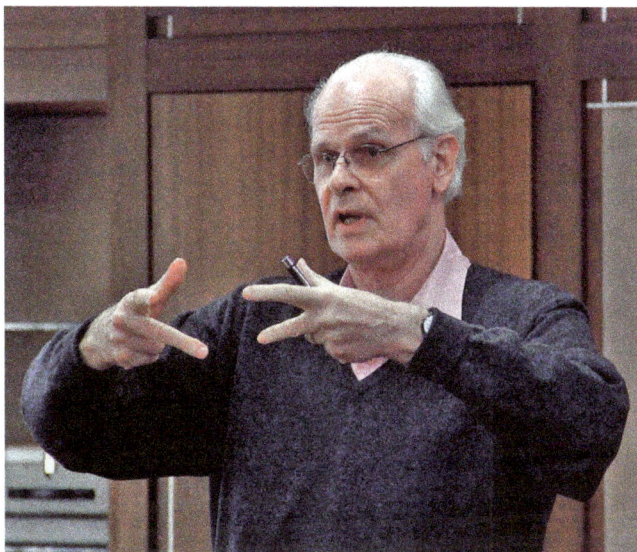

Michel A. Van Hove studied physics in Switzerland (ETH) and the UK (Cambridge), and worked in the Netherlands, Germany, USA, as well as Hong Kong, where he retired as Emeritus Chair Professor. His research has focused on the determination of the atomic-scale structure and bonding at solid surfaces and nanostructures. He developed and implemented novel methods of electron scattering theory and computation, such as for nanostructures. He also worked on photoelectron diffraction, scanning tunnelling microscopy and atomic-force microscopy, a database of solved surface structures, and molecular machines. He published over 400 articles and 12 books. He taught undergraduate and graduate courses in the USA and Hong Kong. This book grew out of a popular general education course which he developed in Hong Kong on "Everyday Physics for Future Executives", inspired by a University of California-Berkeley course on "Physics for Future Presidents".

www.ingramcontent.com/pod-product-compliance
Lightning Source LLC
Chambersburg PA
CBHW061238220326

41599CB00028B/5471